FATE AND TRANSPORT OF NUTRIENTS IN GROUNDWATER AND SURFACE WATER IN AN URBAN SLUM CATCHMENT, KAMPALA, UGANDA

FATE AND TRANSPORT OF NUTRIENTS IN GROUNDWATER AND SURFACE WATER IN AN URBAN SLUM CATCHMENT, KAMPALA, UGANDA

DISSERTATION

Submitted in fulfillment of the requirements of
the Board for Doctorates of Delft University of Technology
and of the Academic Board of the UNESCO-IHE
Institute for Water Education
for the Degree of DOCTOR
to be defended in public on
Monday, 15 September 2014 at 15:00 hrs
in Delft, The Netherlands

by

Philip Mayanja NYENJE

Master of Science in Water Resources Engineering,
Katholieke Universiteit Leuven and Vrije Universiteit Brussels, Belgium

born in Kampala, Uganda

This dissertation has been approved by the promotor:
Prof. dr. S. Uhlenbrook

Composition of Doctoral Committee:

Chairman	Rector Magnificus TU Delft
Vice-Chairman	Rector UNESCO-IHE
Prof. dr. S. Uhlenbrook	UNESCO-IHE / Delft University of Technology, promotor
Dr. J.W.A. Foppen	UNESCO-IHE, copromotor
Prof. dr. J. Griffioen	Utrecht University
Prof. dr. F. Kansiime	Makerere University, Uganda
Prof. dr.ir. P.N.L. Lens	UNESCO-IHE / Wageningen University
Prof. dr.ir. N.C. van de Giesen	Delft University of Technology
Prof. dr. D. Brdjanovic	TU Delft / UNESCO-IHE, reserve member

CRC Press/Balkema is an imprint of the Taylor & Francis Group, an informa business

© 2014, Philip Nyenje

Published by:
CRC Press/Balkema
PO Box 11320, 2301 EH Leiden, The Netherlands
e-mail: Pub.NL@taylorandfrancis.com
www.crcpress.com – www.taylorandfrancis.com

ISBN 978-1-138-02715-2 (Taylor & Francis Group)

Acknowledgements

Many people and organizations have made this PhD study possible. I thank them all very much for their contributions and support.

First, I would like to express my deepest gratitude to my promoter prof. dr. Stefan Uhlenbrook and my supervisor Dr. Jan Willem Foppen for guiding me throughout this research. It has been a long journey and I thank you for keeping the faith in me. This work was largely shaped by my supervisor Dr. Foppen, who from the first day worked hard to change my mindset from modeling to processes understanding and data collection, which was crucial for my study. I cannot count the number of email exchanges and discussions we have had. Thank you for tirelessly reviewing my draft papers that formed the body of this thesis and for all the criticisms that have helped to improve the quality of this research. I am grateful to my local supervisors from Makerere University, Dr. Robinah Kulabako and Dr. Andrew Muwanga, for all their guidance and advice. Dr. Kulabako, thank you for always reminding me to be critical and assertive to ensure that activities were done. Dr. Muwanga, I still have the text books and articles you gave me. They have been very useful especially towards the end of my research when everything started falling into place. I also want to acknowledge Prof. William Paul Johnson (University of Utah, USA) for assisting me during the initial conceptualization of the study and during the initial installation of the monitoring network in Bwaise III slum.

To the entire staff of UNESCO-IHE laboratory, I greatly acknowledge your assistance during the laboratory analyses in Delft. Particularly, I wish to thank Fred Kruis, Don Van Galen, Lyzette Robbemont, Frank Wiegmen and Peter Heerings. Lyzette and Don, thank you for all arrangements in procuring and sending research equipment to Uganda. At Makerere University Public Health and Environment Engineering laboratory, I am particularly grateful to Rita Nakazibwe, Joel Kinobe and John Omara for all their assistance with laboratory analyses of water samples. I also acknowledge the assistance from Bonny Balikkuddembe and Fred Mukasa during soil tests and analyses. In the field, I worked with many people but specifically I want to acknowledge Mr. Dirisa Diodi who was always available whenever I needed him. In this aspect, I also extend my sincere gratitude to the residents in Bwaise III slum for being cooperative, for allowing me install monitoring wells in their homes and for looking after my equipment and the divers. Thank you very much.

To my colleagues Alex Katukiza (now Dr.) and John Bosco Isunju with whom I started this research, thank you for all the moral support and the good times we had. I am grateful to my other PhD colleagues in the Netherlands and other parts of the World with whom we shared experiences: Hans Komakech (now Dr.), George Lutterodt (now Dr.), Heyddy Calderon, Omar Munyaneza (now Dr.), Chol Abel (now Dr.), Girma Ebrahim (now Dr.), Sirak Gebrekristos, John Wasige (now Dr.), Ronald Musenze, Kenan Okurut, Ann Nakagiri, Peter Mutai and Swaib Semiyaga and many more. I want to acknowledge Lisa Meijer and Jasper Havik whose MSc studies contributed significantly to my study. I hope you enjoyed your time in Uganda. The contributions from Nicholas Matsiko and James Lwangwa are also acknowledged.

My stay in the Netherlands was a home away from home. I enjoyed running with coach Jeltsje Kemerink and the running team in the country side of Delft and I always remember the medals we won during our competitions. I cannot forget football at TUDelft with coaches Klaas Schwartz and Davide Merli. I also remember the cold and lonely times during Christmas holidays when I was hosted by Rev. Waltraut Stroh. Thank you very much. I enjoyed the Dutch life, food and parties during the short time I stayed with Leonie Zweekhorst. Thank you for welcoming me in your home and introducing me to your lovely family. I am also grateful to the members of the International Student Chaplaincy Delft and especially to Rev. Waltraut Stroh for the warm fellowship and worship during my stay in Delft.

I am extremely grateful to the Netherlands Ministry of Development Cooperation (DGIS) through the UNESCO-IHE Partnership Research Fund (UPaRF) for funding this research and my travels to and from the Netherlands. Under the initiative of my supervisor, Dr. Foppen, additional funding for the research was also obtained from UPaRF through small-size follow-up research projects.

I extend my gratitude to the entire project team of the SCUSA project (Integrated approaches to address Sanitation Crisis in Unsewered Slum Areas in African Mega cities) for their contributions during our meetings, presentations and interactions. Particularly, I am grateful to the project manager, dr. Jan Willem Foppen and our local team leader, prof. dr. Frank Kansiime of Makerere University. Contributions from prof. dr. Piet Lens and dr. Mariska Ronteltap (UNESCO-IHE), and dr. Charles Niwagaba (Makerere University) are also greatly acknowledged. Jolanda Boots (UNESCO-IHE) is acknowledged for all the administrative and financial aspects, and for ensuring the timely transfer of funds to Uganda.

To my wife Violet, my son Samuel and daughter Abigail thank you for enduring the times I have been away and for being supportive and inspiring. To my brothers and sisters thank you for all the encouragement. Special thanks go my parents especially my mother who struggled to educate me and has been encouraging me up to now. I really dedicate this thesis to you.

I would like to acknowledge all who supported me whose names I have not mentioned. It is difficult to exhaust everyone but I thank you all very much. Finally, I give glory to God who gives me strength to persevere even in difficult times.

Philip M. NYENJE

Delft, the Netherlands

Summary

Rapid urbanization, poor planning and lack of financial resources have led to widespread development of informal settlements (or slums) in urban areas in sub-Saharan Africa (SSA). These areas are usually unsewered and lack access to proper on-site sanitation systems. This often results in the disposal of untreated or partially treated wastewater into the environment, hence contaminating groundwater and surface water. A major consequence of this contamination is the introduction of wastewater-derived nutrients to groundwater and surface water. In excess, nutrients (particularly nitrogen, N and phosphorus, P) can impair the water quality of surface water bodies due to eutrophication. Many fresh water bodies in urban areas of sub-Saharan Africa are indeed increasingly becoming eutrophic, primarily due to the excessive discharge of nutrients from urban informal settlements. There is, however, limited knowledge on the processes governing the transport of these nutrients in urban informal settlements. Hence, the aim of this thesis was to identify the dominant hydrochemical and geochemical processes governing the transport and fate of sanitation-related nutrients in surface water and groundwater systems in an unsewered urban slum area. Understanding these processes is critical to developing effective strategies to minimize nutrient pollution and improve the long-term water quality of urban water resources.

For our study area, we selected a low-lying urban slum area (Bwaise III parish slum; 0.54 km^2) and its catchment (Lubigi catchment; 65 km^2) in Kampala, Uganda. We focused on understanding the main sources of nutrients (N and P) and the processes (hydro-chemical and geochemical) that regulated their transport in groundwater and surface water. The approach we used combined experimental, modeling and processes-description techniques. During the field experiments, we collected a large set of water quality samples from the shallow groundwater, surface water in drainage channels and precipitation in the study area. These samples were analyzed in the laboratory for hydrochemistry (major cations and anions) and nutrients, in particular nitrate (NO_3^-), ammonium (NH_4^+), orthophosphate (PO_4^{3-}) and total phosphorus (TP). In the upper areas of Lubigi catchment, groundwater is often located in deeply weathered regolith aquifers and it was therefore sampled from springs located in the valleys of the catchment. In the low-lying areas where Bwaise III slum was situated, groundwater is located in an alluvial sandy aquifer and it was therefore sampled using shallow monitoring wells (about 1 - 3 metres below ground level), which we installed in selected study sites of the slum area. We also performed detailed hydro-geological investigations in the alluvial aquifer underlying the slum area to characterize the type of aquifer, the soil properties (cation exchange capacity (CEC), available P, texture, geo-available metals and pH) and groundwater flow dynamics (aquifer hydraulic conductivities and flow direction) to gain insights into how they influenced nutrient transport in groundwater. For surface water, we specifically monitored the discharge in the drainage channels and the hydrochemistry and concentrations of nutrients in these channels during low flow and high flow events. We also analyzed the chemical composition of the channel bed and suspended sediments in the channels to understand how they influenced the transport of nutrients in surface water. The data we collected were generally analyzed using descriptive and multi-variate statistics, piper plots, x-y plots, time series, summary tables and the PHREEQC code to identify the sources and dynamics of nutrients, and the dominant geochemical processes that affected their transport in the studied aquifers or drainage channels. The processes considered mainly included reduction-oxidation (redox), precipitation/dissolution, cation exchange and sorption or surface complexation.

The results indicated that both atmospheric deposition and wastewater leaching from on-site sanitation systems were important sources of nutrients (N and P) in the shallow groundwater in the regolith aquifer. Atmospheric deposition, in particular, contributed to N deposition (in the form of NO_3^- and NH_4^+) owing to the presence of nitrogen-containing acid rains, which we largely attributed to the excessive air pollution from motor vehicles in the study area (Kampala city). Consequently, springs in the catchment contained high concentrations of NO_3^- (up to 2 mmol/L) and low pH values (pH < 5). The high concentrations of nitrate in groundwater occurred because the regolith aquifer was slightly oxic implying that the N species in the form of NH_4^+ in wastewater leachates and precipitation recharge were converted to NO_3^- by nitrification. The low pH values, on the other hand, were mainly attributed to acid rain recharge and the poor buffering capacity of the deeply weathered regolith aquifer. We, however, detected low concentrations of PO_4^{3-} (< 2 μmol/L) in groundwater, which we attributed to the strong adsorption of P to Fe-/Al-oxides in the aquifer material. Fe-/Al- oxides were present in large quantities because of the abundance of laterite in the weathered regolith. Geochemical speciation using the PHREEQC code revealed that groundwater was near saturation with respect to $MnHPO_4$ suggesting that this mineral also regulated the sub-surface transport of P by precipitation. Upon groundwater exfiltration to surface water (mainly as springs in this study), nitrate in groundwater was likely lost by denitrification because surface water in the drainage channels was largely anoxic (Mn-reducing). Hence, surface water always contained low concentrations of NO_3^-.

In the shallow alluvial sandy aquifer where Bwaise III slum was located, different processes occurred. Here, shallow groundwater was anaerobic (Fe-reducing) owing to the high nutrient and organic loading from wastewater leaching from the slum area, the presence of organic matter related to wetland vegetation and the low residence time (about 60 years). Hence, NO_3^- in groundwater flowing from the upper regolith aquifer into the shallow alluvial aquifer was almost 100% removed by denitrification. The shallow groundwater, instead contained high concentrations of NH_4^+ (1- 3 mmol/L). It also contained relatively low concentrations of PO_4^{3-} (average 6 μmol/L). These nutrients therefore originated from wastewater leaching directly into the alluvial aquifer from the poor on-site sanitation systems, and particularly from pit latrines, the dominant form of excreta disposal. These pit latrines, however, also retained a substantial of amount of nutrients. We estimated that about 99% of the P mass input and over 80% of N mass input was retained in the pit latrine and the shallow sub-surface in the immediate vicinity of the latrines. Likewise, when we compared the measured nutrient concentrations of the pit latrine leachates (2.4 mg/L or 26 μmol/L as PO_4^{3-} and 57 mg/L or 3.2 mmol/L as NH_4^+) and in the shallow groundwater, we found that the alluvial aquifer in Bwaise slum also removed up to 75% of P and 30 % of N that leached from pit latrines. The removal of PO_4^{3-} in the aquifer was primarily attributed to the adsorption and co-precipitation of P to calcite precipitates whereas the partial removal of NH_4^+ was attributed to the Anammox process. Hence, the pit latrine-alluvial aquifer system acted as an important sink for nutrients that flowed with groundwater from upgradient areas and those that were generated within the slum area from wastewater infiltrating from pit latrines.

Upon exfiltration, groundwater from both the regolith and the shallow alluvial aquifer seemed not to contribute to nutrients in surface water. Surface water, however, still contained high concentrations of dissolved nutrients (in the form of NH_4^+ and PO_4^{3-}), which were about 16 times the minimum required to cause eutrophication. The likely source of these nutrients was the indiscriminate disposal of wastewater (especially grey water: wastewater from bath, laundry and kitchen) directly in the drainage channels. This is a common practice in the poor urban slum areas in SSA. We found that the transport of phosphorus in the primary channel during low flows was regulated by 1) the adsorption of PO_4^{3-} to calcite precipitates, 2) the adsorption of PO_4^{3-} to Fe-oxides especially during high flows when there was re-suspension of bed sediments and, 3) the deposition of organic P. These processes generally led to the retention of P along the channel bed. During high flows, the results showed that the P retained in the bed sediment was occasionally flushed out of the catchment, which further contributed to nutrient loads to downstream streams. The results also indicated that the bed sediments were P saturated and showed a tendency to release P to the overlying water by desorption and mineralization of organic P. These findings provided useful insights into the processes regulating P transport in surface water and groundwater in urban informal catchments. These processes could be useful in developing process-based water-quality models to aid policy and decision making on strategies to reduce the excessive nutrients exported from urban catchments in SSA.

One implication of our findings is that groundwater in urban slum areas in a given physiographic setting may not be of major concern as regards to sanitation-related nutrient pollution and its effects on eutrophication of surface water bodies. This is because pit latrines (the most common on-site sanitation system in slum areas) and the underlying shallow aquifer system act as a large reservoir for nutrients, especially for P, the limiting nutrient for eutrophication. The use of on-site sanitation systems such as improved pit latrines should therefore be encouraged in such areas as a way of minimizing groundwater nutrient pollution. Our findings also indicated that air pollution from combustion of fossil fuels especially from the increasing number of motor vehicles in cities contributed to nitrate pollution in groundwater. Although this may not necessary have an impact on surface water due to denitrification, there could be immediate health risks when this water is directly consumed from community springs or boreholes. Nutrient pollution management strategies should therefore take into account the need to limit air pollution from motor vehicles. Another important implication of our research is that strategies to improve sanitation for environmental sustainability should focus primarily on managing the nutrients transported in surface water. These strategies should, on the one hand focus on carrying out more process-based studies in order to have a better understanding of the processes that regulate P transport, especially in surface water. Knowledge of these processes can be useful in developing improved process-based water quality models, which can aid policy and decision making. On the other hand, a simple phosphorus management practice could focus on minimizing the direct discharge of wastewater to surface water by installing grey water treatment units at household level especially in slum areas of the urban catchment. This is because grey water was identified as the largest wastewater stream that introduces excessive nutrients to surface water in these catchments. To ensure good ecological status of surface water, the required grey water treatment efficiency was estimated to be at least 90% for P. The impact of these strategies may, however, only be seen after a long time (say a decade) because the channel bed sediments in the catchment were P saturated and will likely continue contributing to P loads in the drainage channels.

Samenvatting

Snelle urbanisatie, slechte planning en een gebrek aan financiële middelen hebben geleid tot het op uitgebreide schaal ontstaan van sloppen in urbane gebieden in het zuidelijk deel van Afrika. Deze gebieden zijn meestal niet gerioleerd en ook zijn er geen of onvoldoende goed ingerichte individuele toilet en sanitatie systemen. Hierdoor komen afvalwater en uitwerpselen in het milieu terecht, waardoor grond- en oppervlaktewater vervuilen. Een van de gevolgen is de introductie van nutrienten in het milieu. Als er van die nutriënten (met name stikstof N en fosfor P) teveel in het oppervlaktewater terecht komt, dan kan eutrofiëring optreden. Veel open water in urbane gebieden in zuidelijk Afrika worden inderdaad in toenemende mate eutroof en dan vooral door de uitspoeling van nutrienten afkomstig van afvalwater uit de sloppen. Er bestaat weinig kennis over de processen, die optreden bij transport van deze nutrienten in de sloppen. Het doel van deze dissertatie is om de dominante hydrochemische en geochemische processen te identificeren, die optreden bij het transport van sanitatie gerelateerde nutrienten in oppervlakte water en grondwater systemen in een niet gerioleerde sloppenwijk. Alleen een beter begrip kan leiden tot de ontwikkeling van effectieve strategieën en manieren om vervuiling door nutriënten te verminderen en om een goede water kwaliteit van urbaan water op de langere termijn te waarborgen.

Onze studie vindt plaats in een relatief laag-gelegen sloppenwijk (Bwaise III Parish; 0.54 km^2), inclusief het stroomgebied waarin de sloppenwijk ligt (Lubigi catchment; 65 km^2). De focus van het werk ligt op het achterhalen van de oorsprong van nutrienten en op de processen, die een rol spelen bij transport door de sloppenwijk. Hiervoor wordt een combinatie van experimentele, modelmatige en meer beschrijvende methoden en technieken gebruikt. In het veld is data verzameld omtrent de chemie van ondiep grondwater, oppervlaktewater (drains en sloten) en hemelwater. Geanalyseerde parameters zijn anionen, kationen en nutrienten nitraat, ammonium, ortho-fosfaat en totaal-P. In het bovenstroomse gebied van Lubigi komt grondwater voor in de diep verweerde regolieten, die hier een lokaal watervoerend pakket vormen en her en der door o.a. bronnetjes ontwaterd worden. In het meer benedenstrooms gebied, waarin ook de Bwaise sloppenwijk ligt, komt grondwater voor in een alluviaal zandig ondiep relatief dun pakket, die via 1-3 m lange peilbuizen, die voor deze studie daarvoor zijn geplaatst, zijn bemonsterd. Op meer detailniveau zijn cation exchange capacity, beschikbaar P, textuur, geo-beschikbare metalen en pH bepaald, inclusief aquifer parameters zoals doorlaatfactor en stromingsrichting teneinde inzicht te krijgen in het effect van deze parameters op nutrient transport. Voor oppervlaktewater is gedurende enkele jaren afvoer bepaald, en ook de hydrochemie inclusief nutrienten transport, zowel in perioden van hoge als lage afvoer. Ook zijn de geochemische samenstelling van drain bodems en zwevend sediment geanalyseerd en hun eventueel effect op nutrienten transport. De verzamelde data is geanalyseerd middels beschrijvende multi-variate statistiek, Piper diagrammen, X-Y diagrammen, tijdsseries en PHREEQC. Dit alles om de oorsprong en dynamiek van nutrienten en optredende processen te identificeren in zowel watervoerende pakketten als ook in oppervlaktewater. Beschouwde proces typen zijn o.a. redoxreacties, neerslagreacties, kationuitwisseling, sorptie en oppervlakte complexatie reacties.

De resultaten geven aan, dat zowel atmosferische depositie en infiltratie van afvalwater naar het ondiepe watervoerend pakket belangrijke bronnen van nutrienten vormen. Met name atmosferische depositie draagt bij aan depositie van stikstof (in de vorm van nitraat en ammonium) middels zure regen met hoge N-gehalten. De oorzaak hiervan ligt volgens ons in de excessieve luchtvervuiling veroorzaakt door de enorme hoeveelheid gemotoriseerd verkeer in Kampala en omstreken. Als gevolg daarvan zijn nitraat gehaltes in bronnen hoog (tot 2 mmol/L) bij lage pH waarden (<5). Hoge concentraties nitraat lijken ook te wijzen op een oxisch aquifer, waarin ammonium afkomstig uit afvalwater en hemelwater wordt omgezet in nitraat door middel van nitrificatie. Echter, de lage pH waarden worden vooral veroorzaakt door zure regen en door het gebrek aan buffer capaciteit van de regoliet aquifer. Verder worden in het grondwater lage concentraties PO_4^{3-} waargenomen (< 2 µmol/L). Dit kan verklaard worden door sterke adsorptie van P aan Fe/Al oxides, die aanwezig zijn in de aquifer middels de aanwezigheid van laterieten in de regoliet. Geochemische speciatie m.b.v. PHREEQC wijst uit, dat het grondwater vrijwel verzadigd is met $MnHPO_4$, wat suggereert, dat dit mineraal het ondergronds transport van P tot op zekere hoogte reguleert. Bij kwel van grondwater naar het oppervlaktewater wordt het in het grondwater aanwezige nitraat gedenitrificeerd, omdat de drains en sloten in de sloppenwijk grotendeels anoxisch zijn (mangaan-reducerend). Oppervlaktewater in de sloppenwijk heeft altijd zeer lage concentraties (tot geen) nitraat.

In het zandige watervoerend pakket van Bwaise gebeuren verschillende processen. Grondwater is hier anaeroob (ijzer reducerend) vanwege de infiltratie van grote hoeveelheden organisch materiaal bevattend afvalwater, de aanwezigheid van sedimentair koolstof, afkomstig van voormalige moerasvegetatie (de sloppenwijk is in feite een voormalig papyrus moeras), in combinatie met de relatief lange verblijftijden in het watervoerend pakket (ongeveer 60 jaar). Hierdoor wordt nitraat in het grondwater, afkomstig van infiltratie bovenstrooms, vrijwel volledig gedenitrificeerd. In plaats van nitraat bevat het ondiepe grondwater ter plaatse van Bwaise hoge concentratie ammonium (1-3 mmol/L). Ook bevat het relatief lage concentraties PO_4^{3-} (gemiddeld 6 µmol/L). Deze nutrienten zijn afkomstig van directe infiltratie van afvalwater uit zgn. latrines, die alom aanwezig zijn in de sloppenwijk. Echter, deze latrines zijn ook in staat om een substantiële hoeveelheid nutrienten vast te houden. We schatten in dat ongeveer 99% van de P, die de latrine in komt in de latrine zelf of in de onmiddellijke ondergrond rondom de latrine wordt vastgehouden en voor N is dat ongeveer 80%. En van de hoeveelheid N en P, die dan toch nog de ondergrond in lekt onder de latrine wordt stroomafwaarts in het watervoerend pakket nog eens 75% van de P weggevangen en 30% van de geïnfiltreerde N. Verwijdering van PO_4^{3-} is hoogstwaarschijnlijk door adsorptie en co-precipitatie met calciet en de verwijdering van ammonium is waarschijnlijk door anammox. Met andere woorden, het systeem latrine-aquifer blijkt dus een belangrijke opslag functie te hebben voor nutrienten.

Opkwellend grondwater van zowel de regoliet als het ondiepe alluviale watervoerend pakketje onder Bwaise sloppenwijk levert dus geen belangrijke bijdrage aan nutrienten in het oppervlakte water. Echter, nutrienten concentraties in het oppervlaktewater (vooral ammonium en PO_4^{3-}) zijn nog steeds ongeveer 16 x meer dan wat minimaal nodig is om eutrofiering te veroorzaken. De meest waarschijnlijke oorzaak hiervan is de toevoeging van 'grijs water' aan het oppervlakte water systeem, afkomstig van de 'badkamer', de keuken en het wassen van kleren. Het is in de sloppen van Kampala heel gebruikelijk om dit water direct in de sloten en drains weg te gooien. Als het daar in die drains is, dan blijkt dat tijdens lage drainafvoer, het transport van P in de drains gereguleerd wordt door 1) adsorptie van PO_4^{3-} aan neerslag van calciet, 2) adsorptie van P aan ijzer-oxiden en 3) depositie van organisch P. Tijdens perioden van hoge afvoer werd P weggespoeld uit de drains het stroomgebied uit. Verder wijzen de resultaten uit, dat de ondiepe sedimenten in de drain bodems P verzadigd zijn en de neiging hebben om P af te staan aan het oppervlaktewater door desorptie en mineralisatie van organisch P.

Deze resultaten zijn een zinvolle bijdrage aan de kennis omtrent processen, die van belang zijn bij het transport van P in oppervlakte en grondwater in sloppenwijken. Ook kunnen de resultaten gebruikt worden bij de ontwikkeling van proces gebaseerde water kwaliteits modellen ter ondersteuning van beleid en strategieën om excessieve hoeveelheden nutrienten afkomstig van sloppenwijken in zuidelijk Afrika te reduceren.

Een belangrijk resultaat van dit werk is, dat grondwater in ieder geval in de geografische setting van het onderzoeksgebied niet erg belangrijk lijkt bij afvalwater gerelateerde nutrient problematiek en het effect daarvan op eutrofiering benedenstrooms. Dit is omdat het onderliggende watervoerend pakket zich gedraagt als een opslag en/of afbraakmedium voor nutrienten en dan met name voor P, de meest limiterende nutrient voor eutrofiering. Het gebruik van latrines moet daarom worden aangemoedigd in dit soort gebieden. Onze resultaten wijzen ook uit, dat luchtvervuiling veroorzaakt door gemotoriseerd verkeer in Kampala medeveroorzaker is van nitraatvervuiling in de ondergrond. Alhoewel het effect hiervan op oppervlaktewater uiteindelijk gering is, doordat aeroob grondwater bij kwel gedenitrificeerd wordt, zijn er wel degelijk diverse gezondheidsrisico's voor mensen, die grondwater afkomstig uit de diverse bronnetjes in Kampala direct gebruiken voor consumptie. Strategieën om nutrient vervuiling te voorkomen dienen derhalve de luchtvervuiling mede in ogenschouw te nemen. Een andere belangrijke implicatie van dit werk is, dat strategieën om sanitatie te verbeteren vanuit milieu duurzaamheids oogpunt zich vooral zouden moeten concentreren op nutrienten in oppervlaktewater. Aan de ene kant zouden deze strategieën zich moeten focussen op het krijgen van een nog beter begrip van processen, die van invloed zijn op het transport van P. Aan de andere kant kan een eenvoudige P-beheers maatregel zich concentreren op het verminderen van directe lozing van grijs water in het oppervlaktewater systeem middels de installatie van zgn. grijs-water filter eenheden op het niveau van enkele of meerdere huishoudens. Dit is omdat grijs water de belangrijkste bron van nutrienten in het oppervlaktewater is. Teneinde een goede ecologische status van het water te krijgen moet de efficiency van deze filters wel ongeveer 90% m.b.t. de verwijdering van P zijn. De impact van dergelijke strategieën in het stroomgebied zal echter pas na lange tijd zichtbaar worden, omdat de sloot en drainbodems min of meer fosfaat verzadigd zijn.

List of Symbols and Acronyms

Symbols:

C_{ads}	-	Adsorbed phosphorus	$[M\ M^{-1}]$
CEC	-	Cation exchange capacity	$[mols\ M^{-1}]$
C_{eqm}	-	Equilibrium phosphorus concentration	$[M\ L^{-3}]$
C_{max}	-	Maximum adsorbed phosphorus	$[M\ M^{-1}]$
K	-	Saturated hydraulic conductivity	$[L\ T^{-1}]$
K_{sp}	-	Equilibrium solubility product	$[-]$
IAP	-	Ion activity product	$[-]$
n	-	Porosity	$[-]$
OC	-	Soil organic Carbon	$[M\ M^{-1}]$
OM	-	Soil organic Matter	$[M\ M^{-1}]$
OP	-	Soil organic phosphorus	$[M\ M^{-1}]$
p	-	P – value of statistical significance	$[-]$
pe	-	redox potential	$[V]$
PP	-	Particulate phosphorus	$[M\ L^{-3}]$
Q	-	Discharge	$[L^{3}\ T^{-1}]$
SI	-	Saturation index	$[-]$
SS	-	Suspended solids	$[M\ L^{-3}]$
T	-	Temperature	$[°C]$
TP	-	Total phosphorus	$[M\ L^{-3}]$

Acronyms:

ANAMMOX	-	ANAerobic AMMonium OXidation
HCA	-	Hierarchical cluster analysis
IC	-	Ion chromatography
ICP	-	Ion Coupled Plasma Spectrophotometer
m.b.g.l	-	meters below ground level
PC	-	Principal Component
PCA	-	Principal Component Analysis
PVC	-	Polyvinyl chloride
SSA	-	sub-Saharan Africa
WHO	-	World Health Organization

Units and important conversions:

Units

mg/l	-	milligram per litre
mmol/l or mM	-	millimole per litre
µS/cm	-	micro Siemens per centimeter

Conversions factors for nutrients

From	To	Multiply by
NO_3-N (mg/L)	NO_3^- (mg/L)	4.4
NH_3-N (mg/L)	NH_4^+ (mg/L)	1.288
PO_4 -P (mg/L)	PO_4^{3-} (mg/L)	3.065

Molar conversions

PO_4^{3-} (mg/L)	PO_4^{3-} (mmol/L)	1/95
NO_3^- (mg/L)	NO_3^- (mmol/L)	1/62
NH_4^+ (mg/L)	NH_4^+ (mmol/L)	1/18

Table of contents

Chapter 1

Introduction

1.1. Background

Rapid urban growth, poor planning and management systems and lack of financial resources have led to widespread and almost inevitable development of low-income urban informal settlements or urban slums in (mega-) cities in sub-Saharan Africa (Cronin et al., 2006; Foppen and Kansiime, 2009; Kulabako et al., 2004; Love et al., 2006; Mireri et al., 2007; UN-Habitat, 2003). Sub-Saharan Africa already hosts the largest proportion of urban population residing in slums estimated at 72% in 2001 (UN-Habitat, 2003). Slums are characterized by high population densities, lack of basic services like water and sanitation, poor drainage, lack of secure tenure and insufficient living space (Katukiza et al., 2012; Katukiza et al., 2014; Kulabako et al., 2010; UN-Habitat, 2003). Due to lack of security of tenure, most slums are located in low-lying areas that have been reclaimed from wetlands or swamps (UN-Habitat, 2003). Fig. 1.1 gives a pictorial impression of selected urban slums in sub-Saharan Africa.

Figure 1.1:   A pictorial impression of selected urban slums in sub-Saharan Africa (A) Old Fadama, Accra, Ghana (B) Bwaise III parish, Kampala, Uganda, and (C) Kibera, Nairobi, Kenya (Source: SCUSA research project).

Poor sanitation in slum areas is one of the major environmental concerns in urban catchments. Most often, a sewer system is not present and the commonly-used low cost on-site wastewater handling and reuse practices are frequently unplanned, uncontrolled and inefficient (Cronin et al., 2007; Foppen and Kansiime, 2009; Lawrence et al., 2000). It is reported that over 63% of the urban population in (mega-) cities in sub-Saharan Africa relies on on-site sanitation systems (Nyenje et al., 2010). Hence, most households often poorly dispose off their untreated solid and liquid waste on-site generating high rates of infiltration to aquifers and pollution loads into streams and fresh water bodies (Katukiza et al., 2010a; Kelderman et al., 2009; Kimani-Murage and Ngindu, 2007). This has resulted into excessive release of nutrients (nitrogen, N and phosphorus, P) directly to surface water or via groundwater in most of the growing cities in sub-Saharan Africa (ARGOSS, 2002; Cronin et al., 2006; Kelderman et al., 2009; Nyenje et al., 2010; Xu and Usher, 2006). Excessive nutrients cause eutrophication of surface water bodies leading to a number of environmental

1

problems such as excessive growth of green algae and the water hyacinth, fish kills due to depletion in oxygen levels, release and accumulation of toxic substances and reduced water quality due to anaerobic conditions (Nyenje et al., 2010). Eutrophication also posses direct risks to public health because most large cities may depend entirely on surface water systems for drinking water supply. In the East African region for example, a number of large cities like Kampala in Uganda, Kisumu in Kenya and Mwanza in Tanzania depend on Lake Victoria for their daily water supply. The water quality of this lake has, however, deteriorated over the years due to excessive nutrient discharges from surrounding urban areas (e.g. Mwanuzi et al., 2003; Oguttu et al., 2008) and from atmospheric deposition (e.g. Scheren et al., 2000). With the rapid emergency of slums coupled with destruction of wetlands (or natural filters), most urban catchments will not be able to provide clean water implying that water supply based on surface water systems may soon be unsustainable. This in turn may constraint efforts of achieving the UN millennium goals such as MDG 7 (ensuring environmental sustainability).

The essential nutrients that cause eutrophication are nitrogen (N) and phosphorus (P). However, few studies have investigated the fate and transport of these nutrients (P and N) in groundwater and surface water in urban slum environments. Most studies have focused on direct health risks related to groundwater contamination since groundwater is in most cases the only available source of potable water (Byamukama et al., 2000; Cronin et al., 2007; Howard et al., 2003; Kulabako et al., 2007; Nsubuga et al., 2004; Tredoux and Talma, 2006; Wakida and Lerner, 2005; Zingoni et al., 2005). These studies have indeed found high numbers of coliform bacteria and high levels of nitrate contamination in groundwater, primarily attributed to infiltration of wastewater from unimproved sanitation technologies (e.g. unlined pit latrines) to shallow groundwater and the indiscriminate discharge of grey water (the wastewater from bath, laundry and kitchen) and solid waste directly into drainage channels or over backyard compounds. Despite these studies, sanitation improvement for both public health and environmental protection in slum areas still faces persistent problems and has been less successful. An example is the continual rise in eutrophication of Lakes in urban areas in sub-Saharan Africa caused by uncontrolled discharge of nutrient-rich domestic sewage in these areas (WWAP, 2009).

To properly understand environmental risks and to manage nutrient pollution in groundwater and surface water in poorly sanitized unsewered catchments in SSA, it is necessary to have knowledge of the processes affecting nutrients from the sources where they are released to the sampling points (Mikac et al., 1998; Runkel and Bencala, 1995; Tredoux and Talma, 2006). This is critical to developing strategies and effective policies for improving sanitation in urban informal settlements and reducing degradation of environmental resources.

1.2. Nutrient transport processes

The transport of solutes in groundwater and surface water is influenced by a variety of processes. The most obvious are the physical processes where the solute moves with the water medium by advection and dispersion. Physical processes are more dominant during the transport of conservative or non-reactive solutes such as the chlorides and bromides. For reactive solutes such as the nutrients N and P, other processes may be dominant (Appelo and Postma, 2007; Runkel and Bencala, 1995). These other processes include a number of biological and chemical/geochemical reactions, which can greatly influence solute concentrations in groundwater and surface water. They have not been well characterized in urban informal settlements in sub-Saharan Africa and are therefore the basis of this study, with a focus on the geochemical/hydrochemical processes affecting the fate and transport of nutrients.

1.2.1. Nitrogen

The dominant processes related to nitrogen (N) are nitrification and denitrification (Fig. 1.2). These processes occur depending on the redox state of the environment (Rivett et al., 2008). The redox state may be classified as oxic (with measurable amounts of free oxygen, $O_2 > 1$ mg/L) or anoxic (with depleted/absence of free oxygen but presence of bound oxygen such as NO_3 and SO_4) (Appelo and Postma, 2007). In groundwater studies, deeply anoxic environments are commonly referred to as anaerobic environments meaning total absence of free and bound oxygen (Stuyfzand, 1993; von Sperling and de Lemos Chernicharo, 2005). Aerobic then refers to oxic or sub-oxic environments (Appelo and Postma, 2007).

The dominant form of N from wastewater entering the environment (stream, lake, river, soil, and aquifer) is ammonium, NH_4^+. Some organic nitrogen will also be introduced which may be rapidly transferred into ammonium. These N forms are usually represented as Total Kjeldahl Nitrogen (TKN), which is the sum of organic N, ammonium (NH_4^+) and ammonia (NH_3). Under anaerobic conditions, ammonium (NH_4^+) and ammonia (NH_3) are stable, and when conditions become aerobic (presence of free oxygen), ammonium is rapidly oxidised into nitrate (NO_3^-) via instable nitrite (NO_2^-). This process is called nitrification. Conversely, when a nitrate-rich environment becomes anaerobic, nitrate is reduced to nitrogen gas (N_2), which is stable and ultimately may escape from the aquifer. The process whereby nitrate is converted into nitrogen gas is called denitrification. Another process associated with N is the sorption of ammonium ions (NH_4^+) onto sediments through cation exchange. This may result into release of exchangeable ions such as Ca^{2+} and Mg^{2+} in water, causing additional hardness in water systems. Several authors have indeed found that aquifers contaminated with wastewater contain hard water, which is characterized by high concentrations of Ca^{2+} and Mg^{2+} (Foppen et al., 2008; Lawrence et al., 2000; Navarro and Carbonell, 2007). The partial removal of NH_4^+ by anaerobic ammonium oxidation (*anammox*) has also been reported in anaerobic waters impacted by wastewater discharges from on-site sanitation (e.g. Robertson et al., 2012). Anammox is a bacteria-mediated process whereby aqueous NH_4^+ is converted directly to N_2 gas, which ultimately escapes to the atmosphere. All the above processes are summarized in Fig. 1.2.

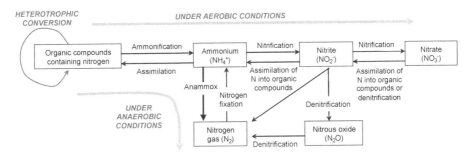

Figure 1.2: Nitrogen transformation processes under different redox conditions (after Lawrence et al., 1997).

1.2.2. Phosphorus

Phosphorus (P) is considered to be the limiting nutrient for eutrophication (Reddy et al., 1999). When phosphorus from wastewater enters the environment, it occurs almost solely as phosphates. Phosphates exist in three forms: inorganic ortho-phosphate, condensed phosphates (pyro-, meta- and other polyphospates), or as organic phosphate (or particulate P) (APHA/AWWA/WEF, 2005; Thornton et al., 1999). Of these forms, ortho-phosphate (PO_4^{3-}) is the most important and readily available form of soluble P, which governs the eutrophication process. The ortho-phosphate ion (PO_4^{3-}) is also simply written as ortho-P or o-PO_4. Organic phosphorus can also be converted to inorganic phosphate during degradation of organic matter through the process called mineralization. The phosphate ion has a strong affinity to adsorb onto soil, aquifer and river sediment grains, and thus, it has a reduced mobility when travelling through soils, rivers and aquifers (Froelich, 1988). The phosphate ion is normally sorbed onto positively charged Fe, Al and Mn oxides and hydroxides (i.e. clay particles) and often tends to accumulate in the soil (Zanini et al., 1998). Orthophosphate is known to have a strong adsorption affinity to Fe oxide or ironoxyhydroxides (FeOOH; common rust) (Golterman, 1995). In hard Ca-rich waters, PO_4^{3-} can also adsorb onto and co-precipitate with calcite (e.g. Bedore et al., 2008; Olli et al., 2009; Golterman, 1995). When soil is eroded, P may erode along with the soil particles and thus be loaded into aquatic systems as sediments. Hence eroded sediment can be a significant source of P loading in water bodies. Besides sorption, phosphates are not very soluble: the solubility products (K_{sp}) of a number of important phosphate salts are very low (Fig. 1.3). As a result, phosphate salts tend to precipitate fairly quickly in order to attain equilibrium. The mineral saturation index (SI) (Eqn. 1.1) is a convenient way of representing the equilibrium condition of a solution with respect to a phosphate mineral/salt. When the calculated SI of a phosphate mineral is close to zero ($SI = 0 \pm 0.5$), then that mineral is usually present and controls the phosphate composition in a water column by precipitation and dissolution (Deutsch, 1997).

$$SI = \log_{10} \frac{IAP}{K_{sp}}$$

(1.1)

Where SI = saturation index (-), IAP = Ionic Activity Product (-) and K_{sp} is the equilibrium constant of a mineral (-).

For $SI = 0$; the mineral is in equilibrium with solution

For $SI < 0$; mineral is undersaturated

For $SI > 0$; the mineral is supersaturated

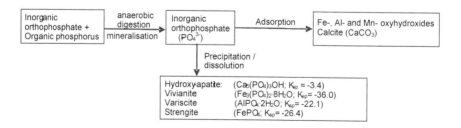

Figure 1.3:   The fate of phosphorus in the environment.

Various authors have also shown that there is a potential for remobilisation of P that has accumulated below sites contaminated with wastewater, which can create dire consequences on the environment. Zurawsky et al. (2004) for example showed that P accumulated below septic tank systems in Ontario in US may be remobilised owing to the reductive dissolution of P present in sediments as $FePO_4$. These reducing conditions can result from the oxidation of dissolved organic carbon, present in wastewater, and transported together with P. Datry et al. (2004) also showed that artificial recharge and rainfall can enhance mineralisation of PO_4^{3-} sorbed on organic sediments resulting in elevated concentration of phosphorus in water. Besides the reductive dissolution of phosphates, which causes remobilisation, P can also desorb from sediments in lake and river bottoms. This is an equilibrium process, which is usually governed by a Freundlich isotherm, whereby the phosphate concentration both on the sediment (P_{sed}) and in surface water (P), plus two constants are important. Based on various research studies, and mostly for surface waters in temperate climates, Golterman and De Oude (1991) arrived at the following relation:

$$P_{sed} = 0.62 P^{0.34}$$

$$(1.2)$$

A consequence of this adsorption mechanism is that after lake restoration measures have been taken, the sediments will release phosphate slowly, thereby delaying the results of the restoration measures. The release of P from surface water sediment can be encouraged by changes in the pH, and it is often believed that the release is also encouraged by decreases in the redox potential (Golterman and De Oude, 1991).

1.3. Problem statement and study objectives

Nutrients, particularly nitrogen (N) and phosphorus (P), are needed for plant growth and healthy ecosystems. In excess, however, they can impair surface water systems giving rise to a range of water quality problems like blooms of algae and the water hyacinth, depletion of oxygen levels and even suffocation or death of aquatic organisms. This can create a number of water supply problems for cities depending on these fresh water bodies due to threats to public health when the affected water body is used for the city's water supply, fishing or recreation purposes. In fact, water supply based on these water resources may in a long-run become unsustainable due to water quality deterioration.

The rapid development of urban informal settlements (or slums) in cities in the Global South due to urbanization has led to high levels of nutrient pollution of waters draining these catchments due to poor on-site sanitation systems. These slums are usually located in low-lying areas and therefore the major wetlands systems that normally filter out nutrients are normally encroached upon and degraded. This implies that nutrients are directly discharged to downstream fresh water lakes hence creating water quality problems due to eutrophication. To properly understand environmental risks and manage nutrient pollution in these water systems, it is necessary to have knowledge of the processes affecting them from the sources where they are released to the discharge points (Mikac et al., 1998; Runkel and Bencala, 1995). However, few studies have investigated the processes governing the fate and transport of nutrients (P and N) in groundwater and surface water in urban slum environments and especially in sub-Saharan Africa, which is characterized by rapid urbanization (e.g. Nyenje et al., 2010). Most studies have focused on understanding immediate health risks resulting from faecal contamination of drinking water supply like springs and shallow wells (e.g. Dzwairo et al., 2006; Howard et al., 2003; Kimani-Murage and Ngindu, 2007). This means that the ultimate fate of nutrients discharged in groundwater and surface water due to poor on-site sanitation systems in urban slum environments is largely unknown. This constrains efforts of safeguarding surface waters from risks associated with eutrophication. It is also difficult to come up with proper strategies for integrated water resources management at catchment level without proper knowledge of the fate of nutrients released into surface water and groundwater systems.

In the context presented above, the overall objective of this thesis was to identify the dominant processes governing the transport and fate of sanitation-related nutrients in surface water and groundwater systems in an urban informal catchment in Kampala Uganda. The focus was on essential nutrients that cause eutrophication, specifically N in the form of nitrate (NO_3^-) and ammonium (NH_4^+) and P in the form of orthophosphate (PO_4^{3-}).

Specifically, the objectives were to:

a) Carry out a critical literature review of eutrophication and nutrient pollution in urban areas in sub-Saharan Africa;

b) Identify the main sources of nutrients (N and P) and pollution pathways/patterns in an urban slum-dominated catchment;

c) Estimate nutrient inputs and related processes from on-site sanitation facilities to shallow groundwater underlying a slum area;

d) Identify the dominant processes governing the fate and transport of dissolved nutrients (NO_3^-, NH_4^+ and PO_4^{3-}) in shallow groundwater and surface water in urban slum areas, and

e) Formulate recommendations for identifying strategies to improve on-site sanitation in order to better manage nutrient pollution and improve the long term water quality status of groundwater and surface water in urban informal settlements.

1.4. Research framework, inter-disciplinary aspects and study location

This PhD study was part of the SCUSA project (Integrated approaches to address Sanitation Crisis in Unsewered Slum Areas in African mega-cities) and it was carried out between 2009 and 2013 (www2.unesco-ihe.org/scusa). The overall aim of the SCUSA project was to seek integrated strategies of improving sanitation in poorly sanitized slum areas in mega-cities in sub-Saharan Africa, taking into account three (3) inter-disciplinary aspects:

(i)   Low-cost sanitation technologies (sanitation component),

(ii)   Socio-economic aspects (socio-economic component), and

(iii)   Environmental impact aspects (hydrology component).

This study formed the hydrology component of the SCUSA project with the aim of assessing the impact of un-sanitized urban slum areas on groundwater and surface water quality. The SCUSA research project was carried out with the collaboration of UNESCO-IHE Institute for Water Education (Delft, The Netherlands), Makerere University in Kampala Uganda and

Kampala City Council (KCC) (now Kampala Capital City Authority, KCCA). KCCA is the municipal authority responsible for providing public services (e.g. health, storm water management, garbage collection and, water and sanitation) in the city of Kampala, the capital city of Uganda.

The project was carried out in Bwaise III parish (32° 33.5'E, 0° 21'N), one of the slums in Kampala, the capital city of Uganda and in Lubigi catchment where Bwaise III slum is located. Lubigi catchment has one of the highest numbers of informal settlements in Kampala city and serves as a good example of a slum-dominated catchment.

This PhD study was carried out at both the catchment scale (Lubigi catchment) and at micro-plot scale (experimental sites in Bwaise III parish slum) using experimental, modeling and a processes-description approach. The later approach, however, formed the major part of our study in order to be able to understand of the dominant processes controlling the fate of nutrients. At catchment scale, the study aimed at obtaining a snapshot of the nutrient generation and transport processes. Then, to provide insights into the hydro-geochemical processes governing nutrient transport and their fate in the shallow groundwater and surface water, micro-plot field investigations were carried out in Katoogo and St. Francis zones of Bwaise III parish slum. At these sites, detailed field investigations were carried out including routine monitoring of water quality, nutrients and discharge of both groundwater and surface water. In the laboratory, several experiments were carried out to characterize the nutrients present in soil and water in the study area and their interactions between soil and water.

1.5. Outline of the thesis

The thesis is organized as a series of seven (7) interconnected chapters that aim to answer each of the specific objectives of this study. Chapters 2 to 6 are based on papers published in international peer-reviewed journals. Therefore, each of these chapters has their own introduction and conclusions. Some degree of repetition therefore occurs in the description of the study area and the methods for water quality analyses.

The thesis starts with Chapter 1, which gives an introduction to the research including the background, research objectives and the study area.

Chapter 2 provides a detailed review of the state-of-art knowledge with regard to eutrophication in sub-Saharan Africa and the transport of nutrients (N and P) from urban settlements to the environment. Research gaps were identified and in the subsequent chapters, they were systematically investigated.

In Chapter 3, hydro-chemical tracers and multi-variate statistics were used to identify the likely sources of nutrients and the pollution patterns at catchment-scale. Here hydrochemical tracers provided evidence that in addition to wastewater infiltrating from slum areas, nitrogen-containing precipitation recharge was also a significant source of nutrients found in groundwater and/or surface water.

In Chapter 4, the extent of nutrient pollution and the resulting processes occurring upon infiltration of wastewater to the shallow groundwater were investigated at two experimental sites in Bwaise III slum: a pit latrine site and a solid waste site.

In Chapter 5, the fate of these nutrients in a shallow alluvial sandy aquifer was evaluated. This was done by identifying the major hydro-geochemical processes along the identified contaminant plume in the shallow groundwater beneath the slum area. The experimental site consisted of two (2) administrative zones in which up to 26 monitoring wells were installed within the slum area and further down-gradient of the slum boundary.

In Chapter 6, the geochemical processes governing the fate of phosphorus in surface water in a slum area were investigated. This was done by carrying out hourly event samplings of nutrients in a channel draining an urban slum catchment during low flows and high flows.

Lastly, chapter 7 provides a synthesis of all the results obtained. Here, recommendations for developing strategies to improve on-site sanitation are also formulated to see how to improve the long term water quality of groundwater and surface water in urban slum areas and downstream ecosystems. Finally, avenues for future research are also suggested here.

Chapter 2

Eutrophication and nutrient release in urban areas of sub-Saharan Africa — A review

Abstract

Eutrophication is an increasing problem in sub-Saharan Africa (SSA), and, as a result, the ecological integrity of surface waters becomes compromised, fish populations become extinct, toxic cyanobacteria blooms are abundant, and oxygen levels reduce. In this review we establish the relationship between eutrophication of fresh inland surface waters in SSA and the release of nutrients in their (mega-) cities. Monitoring reports indicate that the population of (mega-) cities in SSA is rapidly increasing, and so is the total amount of wastewater produced. Of the total amounts produced, at present, less than 30% is treated in sewage treatment plants, while the remainder is disposed of via on-site sanitation systems, eventually discharging their wastewater into groundwater. When related to the urban water balance of a number of SSA cities, the total amount of wastewater produced may be as high as 10–50% of the total precipitation entering these urban areas, which is considerable, especially since in most cases, precipitation is the most important, if not only the 'wastewater diluting agent' present. The most important knowledge gaps include: (1) the fate and transport mechanisms of nutrients (N and P) in soils and aquifers, or, conversely, the soil aquifer treatment characteristics of the regoliths, which cover a large part of SSA, (2) the effect of the episodic and largely uncontrolled removal of nutrients stored at urban surfaces by runoff from precipitation on nutrient budgets in adjacent lakes and rivers draining the urban areas, and (3) the hydrology and hydrogeology within the urban area, including surface water and groundwater flow patterns, transport velocities, dynamics of nutrient transport, and the presence of recharge and discharge areas. In order to make a start with managing this urban population-related eutrophication, many actions are required. As a first step, we suggest to start systematically researching the key areas identified above.

This chapter is based on:
Nyenje, P.M., Foppen, J.W., Uhlenbrook, S., Kulabako, R., and Muwanga, A., 2010, Eutrophication and nutrient release in urban areas of sub-Saharan Africa — A review: Science of the Total Environment, v. 408, p. 447-455.

2.1. Introduction

Eutrophication is one of the most prevalent global problems of our era. It is a process by which lakes, rivers, and coastal waters become increasingly rich in plant biomass as a result of the enhanced input of plant nutrients mainly nitrogen (N) and phosphorus (P) (Golterman and De Oude, 1991). A recent issue of The Water Wheel (Water Research Commission, South Africa; issue September/October 2008) reports that 54% of the lakes/reservoirs in Asia are impaired by eutrophication, in Europe this is 53%, in North America 48%, in South America 41%, and in Africa 28%. In inland sub-Saharan Africa (SSA), there are many documented cases of eutrophication of fresh water resources. Examples include Lake Victoria, which is shared between Uganda, Tanzania, and Kenya (e.g. Cózar et al., 2007; Hecky and Bugenyi, 1992; Hecky et al., 1994; Kansiime and Nalubega, 1999a; Muggide, 1993; Oguttu et al., 2008; Robarts and Southall, 1977; Scheren et al., 2000; Verschuren et al., 2002; Witte et al., 2008), Lake Chivero in Zimbabwe (Jarvis et al., 1982; Magadza, 2003; Moyo and Worster, 1997; Munro, 1966; Nhapi, 2008; Nhapi et al., 2004, 2006; Nhapi and Tirivarombo, 2004), Lake Albert on the boundary between Uganda and Congo (Campbell et al., 2005; Talling, 1963; Talling and Talling, 1965), various fresh water resources in South Africa, like the Zeekoevlei (Das et al., 2009; Das et al., 2008), Rietvlei (Oberholster et al., 2008), and Lake Krugersdrift (Oberholster et al., 2009), rift lakes in Ethiopia (Beyene et al., 2009; Devi et al., 2008; Talling, 1992; Talling and Talling, 1965; Zinabu et al., 2002; Zinabu and Taylor, 1989), or inland delta lakes and fresh water resources in western SSA, like in Cameroon and Nigeria (Arimoro et al., 2007; Kemka et al., 2006).

Most of the nutrients causing eutrophication are reported to originate from agricultural and urban areas (Jarvie et al., 2006; Thornton et al., 1999). In developing countries, like those in SSA, wastewaters from sewage and industries in urban areas, which are often discharged untreated in the environment, are increasingly becoming a major source of nutrients, causing eutrophication of surface water bodies (Bere, 2007; Beyene et al., 2009; Dillion, 1997; Kemka et al., 2006; Kulabako et al., 2004, 2007, 2008; Mladenov et al., 2005; Nhapi et al., 2006; Nhapi and Tirivarombo, 2004; Thornton and Ashton, 1989; Tournoud et al., 2005; Vos and Roos, 2005). This chapter therefore reviews the state of knowledge with regard to N and P transport from urban settlements into the environment. More specifically, this review tries to establish the loads of these nutrients, their transport routes, and the dominant hydrochemical processes along those routes, including the adverse side effects. We shall limit ourselves to inland sub-Saharan Africa, since the rate of development of (mega-) cities in this region has been alarmingly high over the last decade (WWAP, 2009). Attention is also given to urban slums, because they are a major characteristic of many African cities (Kulabako et al., 2004; UN-Habitat, 2003).

2.2. Effects of eutrophication

Before detailing the relationship between urban areas and eutrophication, it is important to first describe the effects of eutrophication in SSA, in order to highlight the importance of the adverse effects of excess nutrients in fresh water resources. The most prominent example is Lake Victoria. This lake has in recent decades undergone a series of profound ecological changes, including strong increases in phytoplankton primary production (Hecky and Bugenyi, 1992; Muggide, 1993), replacement of diatoms by cyanobacteria as the dominant group of planktonic algae (Kling et al., 2001), large scale blooms of the water-hyacinth, and most importantly, the eradication of several species of endemic cichlid fishes. The elimination of cichlid species has been predominantly associated with a Nile perch population explosion, an introduced pescovore (Barel et al., 1985). However, according to Verschuren et al. (2002) and based on evidence from paleolimnological records of lake bottom sediments (Hecky et al., 1994), eutrophication-induced loss of deep water oxygen started in the early 1960s. This may have contributed to the 1980s collapse of indigenous fish stocks starting with the elimination of suitable habitat for certain deep-water cichlids.

A second adverse effect of eutrophication is the rapid growth of phytoplankton species and aquatic macrophytes. In extreme cases, this leads to the development of mono-specific blooms of cyanobacteria (Oberholster et al., 2005; Oberholster et al., 2009). Harmful cyanobacterial blooms are typically characterized by heavy biomass accumulations that often consist of a single or a few species, usually members of the genera Microcystis and Anabaena (Oberholster et al., 2009). Blooms of cyanobacteria in rivers, lakes, and reservoirs disrupt the normal patterns of phytoplankton succession, decrease phytoplankton diversity, and alter virtually all of the interactions between organisms within the aquatic community — from viruses through zooplankton to fish (Figueredo and Giani, 2001). One of the most serious effects of cyanobacterial blooms is the production of harmful secondary metabolites that have serious adverse effects on the health and vitality of humans and animals (Wiegand and Pflugmacher, 2005).

A third effect is the alteration of the ecological integrity of fresh water resources. This may lead to a decline in macroinvertebrate abundance and composition and species richness (Beyene et al., 2009; Oberholster et al., 2008), including fish species (Campbell et al., 2005) and Diptera larvae (Arimoro et al., 2007) or to remarkable physiological adaptations of phytoplankton communities to nutrient variations (Kemka et al., 2009).

Finally, a fourth effect is the total depletion of oxygen. This is associated with the accumulation and decomposition of dead organic matter which consumes oxygen and generates harmful gases such as methane and hydrogen sulphide. When this occurs, many macroinvertebrates and fish species suffocate, while immobile bottom dwelling species can die off completely. In extreme cases, anaerobic conditions ensue, promoting growth of bacteria such as *Clostridium botulinum* that produces toxins deadly to birds, animals and humans. These toxins are also believed to cause gastro-enteritis amongst children (Zilberg, 1966).

2.3. Evidence of the urban areas causing eutrophication

In an important report on the ecology of inland African lakes, Viner et al. (1981) already indicated that by far the most important problems concerning nutrients in Africa are related to urbanization. Their remark concerned deep lakes, like Lake Victoria, Lake Edward, and Lake Turkana (Coulter and Jackson, 1981), shallow lakes, like Lake Bloemhof, Lake Chad, Lake Chilwa, Lake George, Lake Kioga, Lake Naivasha, Lake Ngami, Lake Okavango, Lake Opi, Lesotho Mountain Lakes, and Lake Wuras (Howard-Williams and Ganf, 1981), man-made lakes, like Lake Kariba, Volta Lake, Lake McIlwaine, presently known as Lake Chivero, including various man-made lakes in South Africa (Adeniji et al., 1981), and the rivers contributing to the inflow of these lakes. Although for Lake Victoria, Scheren et al. (2000) reported that atmospheric deposition contributes the largest input of nutrients together accounting for approximately 90% of phosphorus and 94% of nitrogen, comparative studies done by Cózar et al. (2007) between inshore and offshore lake waters indicate stronger eutrophication effects in the inshore areas of Lake Victoria, where nutrient and chlorophyll-a concentrations are markedly higher (Hecky, 1993; Muggide, 1993). In addition, a recent study carried out by the Ministry of Water and Environment of Uganda reported that large urban centres contribute 72% of the pollution loading into Lake Victoria shores compared to 13% by industries and 15% by fishing villages (MWE, 2007). Kansiime et al. (2007) also showed that Lake Victoria's Murchison bay, due to the inflow of wastewater from Kampala (Uganda), has deteriorated in the past decades, as evidenced by an increased loading of nutrients, presently estimated to be 28mg/l NH_4–N from the initial 5.6 mg/l NH_4–N in 1999.

In western SSA, Kemka et al. (2006) showed that Yaounde Municipality Lake in Cameroon is experiencing hypertrophic eutrophication as a result of the inflow of increasing quantities of domestic wastewater from Yaounde city. Already more than 30 years, Marshall and Falconer (1973), Robarts and Southall (1977) and Thornton and Nduku (1982) showed that serious eutrophication in Lake McIlwaine in Zimbabwe (now known as Lake Chivero) was a result of increased sewage effluent. Nhapi and Tirivarombo (2004) demonstrated the eutrophying influence of the Marimba River, discharging into Lake Chivero (Zimbabwe). This river receives treated wastewater from the Crowborough Sewage Treatment Works in Harare. Bere (2007) reported high nutrient concentrations in the Chinyika River, a tributary of the Mazowe River (Zimbabwe), as a result of sewage inflow from the Hatcliffe Sewage Works. The Hartbeespoort Dam, in South Africa, has become a hypertrophic impoundment, predominantly due to the influence of domestic wastewater discharges from the city of Johannesburg and surrounding areas (Allanson and Gieskes, 1961; Thornton, 1989; Thornton and Ashton, 1989). De Villiers (2007) reported increased nutrient input due to anthropogenic activities in the Berg River (South Africa). Oberholster et al. (2009) studied the influence of toxic cyanobacterial blooms on algal populations in Lake Krugersdrift (South Africa). One of the major causes of eutrophication here is the nutrient rich inflow of water from the Modder River, which receives treated domestic and industrial effluent from the city of Bloemfontein. Another example in South Africa is the Hennops River (Oberholster et al., 2008), which receives treated effluent from the Hartbeesfontein Sewage Purification Works, and causes eutrophication in the Rietvlei nature reserve wetland area. The Borkena River in Ethiopia is hypertrophic due to inflow of wastewater from the towns of Dessie and Kombolcha (Beyene et al., 2009). These towns do not possess sewer lines, sewage treatment plants, or proper solid waste disposal sites, and the inflow of nutrient masses is relatively uncontrolled. A similar situation of uncontrolled disposal of wastewater from informal settlements to surface water was reported for the Umtata River in South Africa (Fatoki et al., 2001) and the Orogodo River in Nigeria (Arimoro et al., 2007). The latter receives an uncontrolled inflow of nutrient masses from the towns of Agbor, Owa-Ofie, Ekuma-Abovo and Oyoko, before it ends up in the swamps between Obazagbon-Nugu and the oil rich town of Oben in Edo State, southern Nigeria. Finally, Kulabako et al. (2004, 2007 and 2008) reported on the anthropogenic pollution occurring below Bwaise III Parish, a peri-urban slum area in Kampala (Uganda), and its linkage to uncontrolled discharge of nutrients via groundwater into Lubigi swamp, causing eutrophication of the swamp and surface waters downstream of the swamp. Uncontrolled discharge of nutrients via groundwater was also reported by De Villiers and Malan (1985) for a small urban catchment near Durban, South Africa. In their study, they found out that the high nutrient concentrations found during base flow conditions in drainage channels were above background levels and were linked to leakages from waterborne sewerage.

2.4. Nutrient production and disposal in urban areas

Most of the examples documented in literature above are related to the controlled release of (treated) wastewater into surface water bodies from wastewater treatment plants. Indeed several studies currently indicate that nutrient production in urban areas is more related to wastewater disposal, especially in densely populated areas (Cronin et al., 2003; Kemka et al., 2006; Wakida and Lerner, 2005). This immediately poses a number of important questions: How much wastewater is produced in the larger cities in SSA, which percentage of the wastewater produced is treated, and what are the predominant treatments and (un)controlled disposal mechanisms?

Table 2.1 gives key figures on the water and sanitation coverage in selected mega-cities in Africa. Coverage in this case refers to the number of people with access to safe and adequate drinking water and improved means of sanitation.

Table 2.1:   Water and sanitation coverage in selected mega-cities in sub-Saharan Africa in 1999.

City (Country)	Pop. (1000s)	Population served — Water (%)	Population served — Sanitation (%)	Water production (l/c/d)	Un-accounted for water (%)	Water Connec-tions (%)	Sewer Connec-tions (%)
East Africa:							
Addis Ababa (Eth)	2,444	98	NMV	40	40	4	NMV
Nairobi (Ken)	2,086	100	99	189	40	78	30
Kigali (Rwa)	445	NMV	NMV	118	-	NMV	NMV
Dar-es-Salaam (Tan)	3,000	61	98	150	60	7.3	5
Kampala (Uga)	1,200	72*	78*	110	32*	71*	7*
Southern Africa:							
Maputo (Moz)	967	99	96	133	34	22	25
Windhoek (Nam)	271	100	100	214	11	83	83
Harare (Zim)	2,380	NMV	NMV	156	30	NMV	NMV
Lusaka (Zam)	1,212	81	NMV	225	56	26	NMV
Mbabane (Swa)	94	75	97	100	32	38	47
Luanda (Ang)	4,000	50	62	30	60	18	17
West Africa:							
Lome (Tog)	806	67	80	66	28	55	1.02
Cotonou (Ben)	667	81	83	62	41	81	0.2
Dakar (Sen)	1,925	78	78	128	26	63	26

Source: (JMP, 1999) and (WHO, 2000); NMV = No meaningful value; * 2007 estimates from Water and Sanitation sector performance report 2007, Uganda.

On average, water supply and sanitation coverage is over 70% across sub-Saharan Africa. However, sewerage coverage is generally below 30%. Sewage treatment plants collect and purify these wastewaters, and dispose the purified product in surface waters. Apparently, in many cases the purification is insufficient, giving rise to eutrophication across SSA, as discussed in the previous section of this review. From Table 2.1, it is clear that the non-sewered part of the total urban population is more than 70%, while 63% (the difference between the percentage 'having access to sanitation' and the percentage 'connected to a sewer') relies on on-site sanitation systems. Therefore, around 63% of the total urban population (minimum of 17% in Windhoek and a maximum of 93% in Dar es Salaam) of the mega-cities mentioned in Table 2.1 rely on either septic tanks or traditional/improved pit latrines, eventually discharging their wastewater into aquifers underlying these urban areas. Based on data in Table 2.1, the potential wastewater flows can be obtained by multiplying the population with the consumption rates factored by the unaccounted for water percentages (Table 2.2). With the exception of a few cities in Southern Africa, such as Windhoek, there is a very high proportion of untreated wastewater across all major cities in Africa compared to the total wastewater production. On average, over 80% of the wastewaters produced in large cities in sub-Saharan Africa are untreated and are either discharged in the soil via on-site sanitation systems or directly discharged into rivers and lakes. Depending on the population size of the city, these untreated wastewater volumes can range from approximately 20 to 60 million m^3/y. When multiplied with the composition of medium strength wastewater (Table 2.3), then an approximate annual average of 1×10^6 kg N and 0.1×10^6 kg P is produced by the mega-cities shown in Tables 2.1 and 2.2.

Table 2.2:   Estimated wastewater volumes in a number of mega-cities in sub-Saharan Africa.

City (Country)	Wastewater production (10^6 m^3/y)	Treated (10^6 m^3/y)	Not treated (10^6 m^3/y)
East Africa:			
Addis Ababa (Eth)	21.4	?	?
Nairobi (Ken)	86.3	25.9	60.4
Kigali	?	?	?
Dar-es-salaam (Tan)	65.7	3.3	62.4
Kampala (Uga)	32.8	2.3	30.5
South Africa:			
Maputo (Moz)	31.0	7.7	23.2
Windhoek (Nam)	18.8	15.6	3.2
Harare (Zim)	94.9	?	?
Lusaka (Zam)	43.8	?	?
Mbabane (Swa)	2.3	1.1	1.2
Luanda (Ang)	17.5	3.0	14.5
West Africa:			
Lome (Tog)	14.0	0.1	13.8
Cotonou (Ben)	8.9	0.0	8.9
Dakar (Sen)	66.6	17.3	49.2

Source: JMP, 1999 and WHO, 2000.

Table 2.3:   Characteristics of low, medium and high strength wastewater.

Parameter	Low	Medium	High
BOD (mg/l)	100	200	350
pH	7	7.2	8
Cl⁻ (mgl/)	10	150	650
Ammonia, NH_4-N (mg/l)	10	25	50
Nitrate, NO_3-N (mg/l)	0	0.2	1.5
t- PO_4 (mg/l)	4	10	36
Alkalinity (mg/l $CaCO_3$)	50	200	400
Na^+ (mg/l)	10	120	460
Ca^{2+} and Mg^{2+} (combined) (mg/l)	5	10	25
Boron (mg/l)	< 0.123 - 2.0		

Source: Feigin et al. (1991)

Of course, the distribution of nutrient load across the urban area differs tremendously. In sewered parts of a town, untreated wastewater recharge rates from leaking sewers to the underlying aquifers are estimated to be relatively low (generally less than 50–100 mm/y; Wolf et al., 2006), while wastewater concentrations are likely to be of low-medium strength. In areas where on-site sanitation systems are used, which is true for the vast majority of the urban population in SSA (63% of the total population, as was calculated above), low wastewater volumes with high concentrations are produced (Table 2.4). In informal settlements, population densities are high, sewerage is lacking, and, if present, sanitation facilities are almost exclusively on-site (mainly pit latrines, ventilated improved pit latrines or VIPs, elevated pit latrines, etc.; Zingoni et al., 2005; Kulabako et al., 2007). The nutrient load produced in these areas is extremely high: the proportion of the urban population residing in slums in SSA is estimated to be 4–80% with an average of around 50% (Figure 2.1b). Given the increasing urbanisation trends in the future (Figure 2.1a), the number of people living in urban slums in SSA is expected to rise continuously. This is expected to give rise to increased concentration and nutrient fluxes from these areas.

Table 2.4:   Faecal sludge characteristics in on-site sanitation systems in Kampala.

Parameters	VIP latrine	Septic tank
Total Solid, TS (mg/l)	30,000	22,000
Total Volatile solids (% TS)	65	45
COD (mg/l)	30,000	10,000
BOD (mg/l)	5,500	1,400
Total Kjedhal Nitrogen, TKN (mg N/l)	3400	1,000
Ammonia, NH_4^+ (mg/l)	2,000	400
Nitrates, NO_3^- (mg N/l)	-	-
Total Phosphorus, TP (mg P/l)	450	150
Feacal coliforms (cfu/100ml)	1×10^5	1×10^5

Source: NWSC (2008)

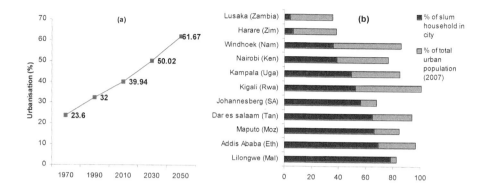

Figure 2.1: Urbanisation trends and slum proportions in Africa: (a) Urbanisation trends %
1970-2050 and, (b) slum proportions (%) in selected cities. The *blue rectangle*
in Fig. 2.1b shows the proportion of the total urban population residing in slums
while the *red rectangle* shows the proportion of slum households in selected
cities (Source: UN-Habitat (2008)).

2.5.   The urban water balance

How important are these loads in the entire water balance of the city and what is the role of
the hydrological processes in the transport of nutrients to surface water bodies? Ideally, a
hypothetical water balance of the upper part of the soil (say from the surface to 2–4 m below
the surface) of an urban area consists of (Figure 2.2): precipitation, evaporation, water
imported/exported, outflow/inflow to/from groundwater, storm water runoff, and sewer
outflow and changes in storages. All these variables indicate flow across the boundary of the
urban area (Marsalek et al., 2008). Within the urban area the following terms can be
discerned: impervious surfaces, soil, household and sewerage (wastewater, storm water, or
combined). As an example, the water balance of Kampala, Uganda (Figure 2.2) is given. The
following dominant fluxes of water (or hydrological pathways) can be identified:

- Most of the precipitation (1450 mm/y) is evaporated (1151 mm/y), while the rest (330
  mm/y - this includes water derived indoor and outdoor uses) flows into Lake Victoria
  via open and closed drains present in Kampala city.

- Around 170 mm/y of water is imported (from Lake Victoria) and used indoor.
  Although there are some leakages (17 mm/y) and outdoor usages (5 mm/y), most of
  this water (148 mm/y) is converted into wastewater, of which 138 mm/y is disposed
  of via on-site treatment facilities (pit latrines, septic tanks, etc.), while 10 mm/y is
  transported to Lake Victoria.

- A part of the wastewater disposed of on-site is largely mixed and diluted with the
  fraction of precipitation infiltrating the soil. Of the total amount of water reaching the
  soil (1245 mm/y), finally around 120 mm/y recharges groundwater, of which only 10
  mm/y reappears as springs. Most of the remainder (1100 mm/y) evaporates, while
  some water (24 mm/y) is stored. Of course, the long term storage component should
  be zero, indicating a situation of steady state. However, in this case, Kampala itself is
  not in a situation of steady state: urbanisation is taking place, and, as one of the

consequences, water is stored in the subsurface, and the groundwater table is — on average — on the rise.

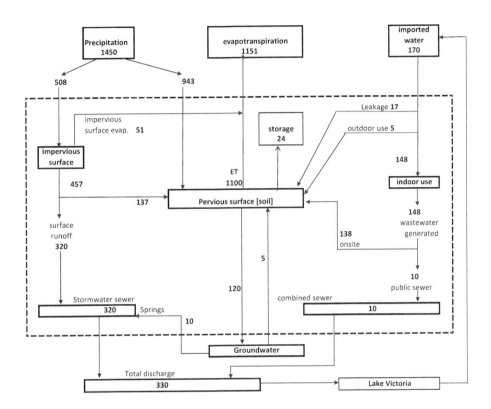

Figure 2.2: Estimated water balance (mm/y) for the upper soil compartment of Kampala city, Uganda. The major components of the water balance are precipitation of 1450 mm/yr (average of Kampala daily precipitation from 2000-2010), evapotranspiration (i.e. total evaporation) of 1100 mm/y (FAO, 1970), imported water of 170 mm/y (estimated from the served population and per-capita production of 110 l/c/d (Kaggwa, 2009; KDMP, 2002)) and groundwater recharge (120 mm/y, Taylor & Howard 1996). Other components include indoor water use (water used inside homes in bathrooms, kitchens, toilets etc), outdoor water use (assumed to be 3% of imported water - includes water used for watering gardens, car washing etc.) and water pipe leakages (assumed to be 10% of imported water). Major assumptions made were: 1) all water consumed is converted to wastewater, 2) all precipitation on pervious surfaces infiltrates the soils, and 3) impervious area (roofs, roads and paved surfaces) is 35% of the total urban area.

Table 2.5 provides an overview of the urban water balance for selected cities in sub-Saharan Africa, including the example of Kampala (Uganda). Based on this table, the amount of wastewater that is disposed of via on-site sanitation facilities or via drainage channels without being treated, ranges from 10% (Keren) to 50% (Khartoum) of the total precipitation entering the urban areas. This is a large percentage of untreated wastewater especially considering the fact that in most cases, precipitation is the most important, if not the only 'wastewater diluting agent' present. It should be mentioned here, that probably most of the wastewater generated is disposed of via on-site sanitation facilities. In recent years, it has even become apparent that the common way of handling wastewater in developing countries through on-site sanitation systems generates rather high rates of infiltration, often referred to as 'modern recharge' in developed countries (Morris et al., 2006; Wolf et al., 2006). Within and outside sub-Saharan Africa, there are many examples testifying of an increased recharge of wastewater (Morris et al., 2003), thereby sometimes even causing local wastewater flooding problems, including associated health problems, road damage, and odour nuisance. The increased recharge due to urbanization is somewhat contradictory, since urbanization implicates the construction of roofs, and paved surfaces, which reduce recharge. Those impervious surfaces do indeed reduce recharge from precipitation, and on many occasions, cause fast surface runoff responses leading to flooding of lower lying parts of the city. Good examples for the latter are the Bwaise III slum area in Kampala (Kulabako et al., 2004, 2007, 2008), which is frequently flooded by storm water from the upland urban catchment area, and various other slum areas in Nairobi (Kenya), Kampala (Uganda), Lagos (Nigeria), Old Fadama in Accra (Ghana), Free Town (Sierra Leone) and Maputo (Mozambique) (Douglas et al., 2008; Monney et al., 2013). In contrast, Kelbe et al. (1991) reported a reduction of the peak discharge from catchments inhabited with informal settlements as compared to similar pristine catchments.

Table 2.5:   Estimates of water balances (in mm/y) of the upper part of the subsurface in selected cities in (Sub-Saharan) Africa.

Component of urban water cycle	Khartoum (Sudan)	Keren (Eritrea)	Sunninghill (South Africa)	Yaounde (Cameroon)	Kampala (Uganda)
Inflow					
Precipitation	140	400	724	1302	1450
Imported water	-	8	114	288	170
River inflow	491	-			
Groundwater inflow (springs)	65	-	-	-	10
Capillary rise	-	-	-	-	5
Outflow					
Evapotranspiration	140	207	457	275	1151
Wastewater flow (sewered)	4	1	95	-	10
Wastewater infiltration	78	5	-	128	120
Recharge to groundwater	390	227	179	-	-
Surface runoff in storm water	-	2	107	1187	330
River outflow	113	-	-	-	-
Change in storage					
Soil storage (vadose zone)	-29	-34	-	-	24

Sources: Kampala: (FAO, 1970; Kaggwa, 2009; KDMP, 2002; Taylor and Howard, 1996); Sunninghill: (Stephenson, 1991). Other cities: estimates provided by expert judgments of relevant ministerial employees.

2.6. Processes related to the nutrients N and P in sub-Saharan Africa

In SSA, research related to processes on N and P has been mainly focused on the natural retention and carrying capacity of the environment. Bere (2007) for example found out that nutrient loads in the Chinyika River (Zimbabwe) were retained over a distance of 4 km from the wastewater treatment discharge point into the river due to a high natural retention capacity of the river. Bere (2007) concluded that the natural retention capacity was most likely the result of the presence of swamps and wetlands near the banks of the river. Wetlands are able to retain nutrients (N and P) through retention by sediments and uptake by plants or by sedimentation of nutrient-rich particulate matter (Kansiime et al., 2007; Kansiime et al., 2005; Kelderman et al., 2007; Mugisha et al., 2007; van Dam et al., 2007). They act as buffers to eutrophication and constitute an important role in maintaining lake and river ecosystems. Van Dam et al. (2007) constructed a dynamic model for nitrogen cycling to understand the processes contributing to nitrogen retention in a wetland and to evaluate the effects of papyrus harvesting on their nitrogen absorption capacity. The model used data from Kirinya wetland in Jinja (Uganda), which receives effluent from a municipal wastewater treatment plant. Kelderman et al. (2007) showed that Kirinya wetland on the Uganda side of Lake Victoria can retain nutrients from secondary treated wastewater up to 40% to 60%. A wetland model was also constructed by Mwanuzi et al. (2003) to simulate the buffering processes of nutrients and organic matter based on results of wetlands in the Tanzanian part of Lake Victoria. The model established that there was a net export of nitrates and organic matter produced in wetlands while most inorganic P (60% to 90%) was retained in wetlands. Kansiime et al. (2005, 2007) showed that papyrus vegetation exhibited a higher wastewater treatment potential than the agricultural crop cocoyam. Mugisha et al. (2007) concluded that wetland plant species with high phytomass productivity and well developed root systems and ability to withstand flooding are best suited for nutrient removal. Encroachment on wetlands and increased wastewater production in urban areas, however, have increased nutrient loading beyond the buffering capacity of wetlands, thus impacting on lake ecosystems. Kansiime et al. (2007) for example found that there was 7 times more nutrient loading in wetlands located in urban areas in Uganda than those in rural areas. This was attributed to urban wastewater discharges with nutrient concentrations reaching as high as 4–7 mg/l of total-P and 15–17 mg/l of total-N. The Nakivubo channel in Kampala Uganda for example currently carries approximately 90% of N and 85% of P, discharged every day into the Inner Murchison Bay of Lake Victoria via Nakivubo wetland (NWSC, 2008).

Although more than 60% of the wastewater in mega-cities is disposed of via the sub-surface, research aimed at identifying the fate and transport mechanisms of both N and P in soils and aquifers is almost completely absent in SSA. An exception is Kulabako et al. (2008) who looked at the fate of P in the Bwaise III slum area in Kampala. They found out that P transport mechanisms are a combination of adsorption, precipitation, leaching from the soil media and by colloids, with the latter two playing a far more important role. These findings concur with previous studies which showed the potential for remobilisation of P accumulated in soils caused by reductive dissolution of P, rainfall recharge, changes in pH and decreases in redox potential (Datry et al., 2004; Golterman and De Oude, 1991; Zurawsky et al., 2004). In addition, the Langmuir and Temkin isotherm equations could be used to describe the P adsorption phenomena of the soils in Bwaise. From sequential and parallel extraction techniques for P present in the soils, residual P consisting mainly of organic phosphorus compounds was the dominant fraction in all soil layers, followed by Ca-bound P, and, finally, Fe and/or Al-bound P. From this, we conclude that these soils, which consist mostly of debris deposited by slum dwellers, possess a natural retention capacity. This capacity is however limited, since P concentrations in shallow groundwater in the area may be as high as 13 mg/l total-P (Kulabako et al., 2004). The capacity to remove nitrogen in these soils seemed to be limited: maximum concentrations of up to 779 mg/l NO_3^- were measured with average concentrations of several 10s of mg/l. In general, all samples taken from the shallow groundwater of the Bwaise III area pointed towards oxic conditions. The presence of nitrate and therefore the oxic state of groundwater is reported for various locations in the region (Alagbe, 2006; Arimoro et al., 2007; Barrett et al., 1999a; Cronin et al., 2007; Dzwairo et al., 2006; Edet and Okereke, 2005; Efe, 2005; e.g. Faillat, 1990; Gelinas et al., 1996; Ikem et al., 2002; Nevondo and Cloete, 1999; Uma, 1993; Yidana et al., 2008; Zingoni et al., 2005). Several researchers in SSA also report on the oxic state of unpolluted groundwater without nitrate (e.g. Diop and Tijani, 2008; Kortatsi, 2006; Mkandawire, 2008; Pritchard et al., 2007, 2008). To our knowledge, there are very few reports testifying of deeply anaerobic conditions (without free oxygen) of urban groundwater, whereby sulphides are produced or methane is formed. Exceptions are discussed in Kortatsi (2007) and Kortatsi et al. (2008), who have reported on the reductive dissolution of iron oxides and the presence of Fe^{2+} in the regolith of Ghana at depths varying between 90 and 120m. Vis-a-vis the 'redox'-model, which was formulated by Lawrence et al. (2000), a supposed lack of (deeply) anaerobic groundwater is somewhat remarkable.

Based on their findings in Hat Yai, a town in Thailand, these authors have stated that dissolved organic carbon (DOC) can be considered the primary driver of chemical processes in aquifers contaminated with wastewater. Decomposition of DOC by bacteria depletes oxygen and causes suboxic and anoxic groundwater. If the decomposition of DOC can continue, then nitrate reduction may take place, followed by the release of Mn^{2+} and Fe^{2+} (i.e. the manganogenic and ferrogenic zones), the production of sulphides (sulphidogenic zone) and, finally, methane (methanogenic zone). It should be noted that the reductive dissolution of Fe oxyhydroxides can result in mobilisation of various sorbed harmful components, like phosphates, arsenic, selenium, and a multitude of polar organic contaminants. It is also worth noting that these components and most especially the exchangeable ions Ca^{2+} and Mg^{2+} can also be released into water through cation exchange processes with ammonium (NH_4^+) in wastewater (Foppen et al., 2008; Navarro and Carbonell, 2007). In the Hat Yai case, all the anoxic zones described above were observed. In all reported cases for urban areas in SSA, not even the first zone (nitrate reducing zone) was reported. This might be due to the following:

- If anaerobic groundwaters exist, then the procedures for anaerobic sample taking were not followed properly.

- Oxygen entry into the unsaturated zone for accepting the electrons required to fulfill the redox reaction is almost unlimited. This might be due to variations in seasonal recharge and the thickness of the unsaturated zone, which may cause seasonal replenishment of oxygen supply in the unsaturated zone.

- There is not enough reactive DOC in wastewaters present in SSA to complete all redox reactions.

- The model suggested by Lawrence et al. (2000) might require reactive sedimentary organic material, which might have been present in the organic rich Hat Yai aquifers, but is most likely almost absent in the weathered regoliths of large parts of SSA.

2.7. Knowledge gaps

In sub-Saharan Africa, eutrophication of fresh water resources, like lakes and rivers, is currently on the rise and most lakes and fresh water sources located near urban areas are deteriorating at an alarming rate. A large part of the problem is caused by the rapid increase in population and urbanisation, especially in informal settlements where there is uncontrolled disposal of wastewater. With over 80% of the wastewater generated remaining untreated, and disposed of in the soil via on-site sanitation systems or directly discharged into rivers, one of the most important questions is "where do the nutrients in wastewater end up?" This question can be sub-divided in a number of more specific research topics:

1. Although most of the on-site facilities dispose off their wastewater in the subsurface, the fate and transport of N and P originating from this infiltrating wastewater is unknown. There is evidence that in wetlands and surface waters denitrification is occurring, resulting in a net loss of nitrogen, which, in turn, could give rise to N being the limiting factor for primary production and eutrophication. The rate of denitrification in aquifers, however, is unknown. The reports on chemical groundwater quality in urban areas in the region suggest that denitrification in aquifers is relatively limited, and that transformation of nitrogen species (e.g. nitrate) is not a dominant process.

2. Phosphorus transport, including sorption, dissolution and precipitation of iron- and manganese-oxyhydroxides and iron phosphate minerals to either lake bottom sediment, riverine sediments or to aquifer materials, and their dynamics as a result of input variations and temperature changes, has not been studied in great detail.

3. The effect of episodic removal of nutrients stored at the urban surface and near surface erosion by runoff from precipitation on nutrient budgets in adjacent lakes and rivers draining the urban areas is unknown. How is this temporary storage of nutrients related to eutrophication in downstream surface waters, and is the temporary influx of nutrients in any way related to the cyanobacteria blooms, which occur frequently in the region?

4. There is no insight in the hydrochemical state of both surface and subsurface contaminated water. Most reports check for WHO guidelines, and/or check for the presence of nutrients/microbiological parameters, but research identifying hydrochemical processes is lacking: what are the dominant redox processes, which dissolution and precipitation reactions take place, is cation exchange an important mechanism, and how important is sorption of contaminants? The processes are widely unknown in most catchments in SSA (mega-) cities.

5. There is very little insight in the hydrological cycle within the urban area, including surface water and groundwater flow patterns and interactions, associated transport velocities, dynamics of nutrient transport, and the presence of recharge and discharge areas in the urban area and their space and time variability in different seasons.

6. Although not specifically addressed in this review, there is no insight in the presence and spreading of organic micro-pollutants in city catchments in SSA. Prominent examples of emerging contaminants are pharmaceuticals, estrogens, ingredients of personal care products, biocides, flame retardants, benzothiazoles, benzotriazoles or perfluorinated compounds (PFC). Adverse effects by individual emerging contaminants can occur with concentrations even as low as a few nanograms per litre, as reported for 17α-ethinylestradiol and tributyltin. Besides endocrine disrupters also pharmaceuticals (e.g. carbamazepine, diclofenac, fluoxetine, and propranolol) have been shown to cause effects at environmentally relevant concentrations. Contaminations of groundwater and drinking

water by emerging contaminants are well reported. Since current treatment processes of municipal wastewater and drinking water treatment plants (e.g. nitrification/denitrification, flocculation and filtration) are not able to remove the majority of these emerging contaminants and advanced treatment techniques (e.g. ozonation, PAC addition and membrane treatment) may either be too costly or do not guarantee complete removal of these compounds, the question that arises is: which are the most hazardous or "unwanted" emerging contaminants? Criteria for answering this question might be the eco-toxicological (aquatic and terrestrial) and toxicological relevance, the potential to bioaccumulate as well as the potential to contaminate groundwater and drinking water. In the SSA region, research on the presence and dynamics of these pollutants in the urban environment, both in groundwater and in surface waters and their linkage to eutrophication is absent.

2.8.   Conclusions

The most important conclusions from this review are:

1. Eutrophication is an increasing problem in sub-Saharan Africa (SSA), and, as a result, the ecological integrity of surface waters becomes compromised, entire populations of fish become extinct, toxic cyanobacteria blooms are abundant, and oxygen levels become depleted, thereby promoting the growth of pathogenic bacteria such as C. botulinum.

2. In the literature, there are many reports establishing the relation between eutrophication of fresh inland surface waters in SSA and the production of nutrients in the various (mega-) cities, which is fundamentally different from eutrophication mainly caused by agriculture in the so-called North.

3. Monitoring reports indicate that the population of these (mega-) cities is rapidly increasing, and so is the total amount of wastewater produced. Of those total amounts produced, at present, less than 30% is treated in sewage treatment plants, approximately 60% (on average) is disposed of untreated via on-site treatment systems, discharging their wastewater eventually into groundwater, while the fate of the remaining portion of total wastewater produced is to a large extent unknown.

4. The most important knowledge gaps include: (1) the fate and transport mechanisms of both N and P in soils and aquifers, or, conversely, the soil aquifer treatment characteristics of the regoliths, which cover a large part of SSA, (2) the effect of the episodic and largely uncontrolled removal of nutrients stored at the urban surface due to runoff from precipitation on nutrient budgets in adjacent lakes and rivers draining the urban areas, (3) the hydrology and hydrogeology within the urban area, including surface water and groundwater flow patterns, transport velocities, dynamics of nutrient transport, and the presence of recharge and discharge areas and, (4) the presence and spreading of other compounds present in wastewater, like organic micro-pollutants, trace metals and microbiological pollutants (viruses, bacteria, and protozoa).

In order to make a start with managing this urban population related eutrophication, many actions are required. As a first step, we suggest to start systematically researching the key areas identified above. In this thesis, we attempted to address some of these research gaps by investigating the sources of nutrients and the processes governing their transport in an urban slum environment in Kampala, Uganda.

Chapter 3

Using hydrochemical tracers to identify sources of nutrients in unsewered urban catchments

Abstract

In this study, we investigated the principal sources of nutrients in an urban-slum dominated catchment. To this, we applied graphical methods and multivariate statistics to understand impacts of an unsewered slum catchment on nutrients and hydrochemistry of groundwater in Kampala, Uganda. Data was collected from 56 springs (groundwater), 22 surface water sites and 13 rain samples. Groundwater was acidic and dominated by Na^+, Cl^- and NO_3^- ions. These ions were strongly correlated indicating pollution originating from wastewater infiltration from on-site sanitation systems. Results also showed that rain, which was acidic, impacted on groundwater chemistry. Using a Q-mode hierarchical cluster analysis, we identified three distinct water quality groups. The first group had springs dominated by Ca-Cl-NO_3, low values of EC, pH and cations, and relatively high NO_3^- concentrations. These springs were shown to have originated from the acidic rains because their chemistry closely corresponded to ion concentrations that would occur from rainfall recharge, which was around 3.3 times concentrated by evaporation. The second group had springs dominated by Na-K-Cl-NO_3 and Ca-Cl-NO_3, low pH but with higher values of EC, NO_3^- and cations. We interpreted these as groundwater affected by both acid rain and infiltration of wastewater from urban areas. The third group had the highest EC values (average of 688 µS/cm), low pH and very high concentrations of NO_3^- (average of 2.15 mmol/L) and cations. Since these springs were all located in slum areas, we interpreted them as groundwater affected by infiltration of wastewater from poorly sanitized slum areas. Surface water was slightly reducing and eutrophic due to wastewater effluents, but the contribution of groundwater to nutrients in surface water was minimal because PO_4^{3-} was absent whereas NO_3^- was lost by denitrification. Our findings suggest that groundwater chemistry in the catchment is strongly influenced by anthropogenic inputs derived from nitrogen-containing rains and domestic wastewater.

This chapter is based on:
Nyenje, P.M., Foppen, J.W., Uhlenbrook, S., and Lutterodt, G., 2013, Using hydrochemical tracers to assess impacts of unsewered urban catchments on hydrochemistry and nutrients in groundwater: Hydrological Processes, DOI: 10.1002/hyp.10070.

3.1. Introduction

Unsewered low-income urban catchments in sub-Saharan Africa (SSA) are characterized by poor on-site sanitation activities and are becoming a major threat to the environment (Dzwairo et al., 2006; Nhapi et al., 2006; Nyenje et al., 2010). This is because wastewater discharged from poor on-site sanitation systems can lead to excessive nutrient loads (phosphorus, P and nitrogen, N) to groundwater and surface water draining these areas giving rise to eutrophication of downstream aquatic systems (Kemka et al., 2006; Kulabako et al., 2007; Nhapi et al., 2006; Nyenje et al., 2010). Eutrophication causes water degradation, which can interfere with various water uses such as water supply and recreation. This is a growing problem that has received little attention in developing countries, where more immediate problems such as public health and settlement are prioritized due to limited funds. However, nutrient rich waters can also cause public health problems like the blue baby syndrome in infants caused by consumption of nitrate-rich groundwater (> 50 mg/L as NO_3^-) (Thornton et al., 1999). A recent review by Brender et al. (2013) also noted that drinking water with high nitrate can cause several birth effects in new borne babies. Nutrients can be released from multiple sources such as atmospheric deposition, leachates from on-site sanitation systems and agricultural activities, and decomposition of organic matter in the soil (Scheren et al., 2000; Wakida and Lerner, 2005). Sewage discharges from on-site sanitation are, however, becoming major sources of nutrients impacting water resources in urban areas in SSA (e.g. Nhapi and Tirivarombo, 2004; Nyenje et al., 2010). Understanding the sources of these nutrients and the dynamics controlling their transport is, however, complex, because of the large number of environmental, social, and economic activities which take place in urban environments. Moreover, the processes that influence the transport of nutrients in unsewered urban catchment in SSA have not been studied in detail (see Chapter 2).

Hydro-chemical tracers have been used successively in the past to understand flow paths, water quality types, origin and variation of hydro-chemical parameters, and processes affecting transport of solutes in aquatic systems (e.g. Didszun and Uhlenbrook, 2008; Lambrakis et al., 2004; O'Shea and Jankowski, 2006; Yidana et al., 2011). However, mere interpretation of hydro-chemical data can produce conflicting signals, thus statistical methods provide more objective data interpretation (Capell et al., 2011; Cloutier et al., 2008). Multivariate statistical techniques are widely used to analyze and interpret large sets of hydro-chemical data. The most common techniques are cluster analysis to group observations with similar hydrochemical characteristics and principal component analysis (PCA) to identify relationships between hydro-chemical variables (Capell et al., 2011; Froelich, 1988; Güler and Thyne, 2004; Güler et al., 2002; Lambrakis et al., 2004). In some studies where only factor analysis is applied, the common factor analysis is preferred to PCA because it is able to explain correlations among variables (Costello and Osborne, 2005). PCA identifies relationships between variables by reducing them to a smaller number of independent principal components taking into account all the variance in the data. The common factor analysis technique is similar to PCA, but it analyzes only the reliable common variance of data, and common factors are estimated basing on the underlying latent structure of the variables (Costello and Osborne, 2005). PCA is, however, most commonly used in statistical programs and in understanding hydrochemical processes (Belkhiri et al., 2010; Costello and Osborne, 2005). Cluster analysis and PCA techniques have been applied successfully in the past to understand anthropogenic processes that control groundwater chemistry. Salifu et al. (2012) for example recently applied PCA on 357 groundwater samples from Northern Ghana to provide an insight into the hydrogeochemical characteristics and the processes influencing the occurrence of fluoride in groundwater. Kebede et al. (2005) used hierarchical cluster analysis (HCA) to classify groundwater into objective water quality groups in order to get an insight into groundwater recharge, circulation and its hydro-chemical evolution. Yidana et al. (2011; 2008) has used HCA and PCA to define recharge processes and the origin of chemical parameters of groundwater in the Upper Volta Basin system in Ghana. Güler et al. (2012) also used cluster analysis and PCA to assess the impact of anthropogenic activities on groundwater hydrology and chemistry and to identify hydro-geochemical processes occurring in the Tarsus coastal plain in South Eastern Turkey. Few studies have, however, been carried out in unsewered slums areas in growing cities in SSA.

In this study, we applied graphical methods and multivariate statistics to gain insights into the impacts of an urban slum-dominated catchment on nutrients and hydrochemistry of groundwater. Here, slums refer to unplanned informal settlements lacking proper sanitation. The catchment was located in Kampala, the capital city of Uganda. More specifically, our objectives were to (1) identify sources of dissolved nutrients NO_3^-, NH_4^+ and PO_4^{3-}, and (2) understand dominant mechanisms controlling the transport of these nutrients.

3.2.   Study Area (Lubigi catchment)

3.2.1. Location and land use

The Lubigi catchment is an ungauged catchment of 65 km^2 located Northwest of Kampala, the capital city of Uganda (Fig. 3.1). It has an estimated population of about 392,924 of which the majority has low income estimated at less than 2 USD per person per day (CIDI, 2006; KDMP, 2002). A large part of the catchment is urbanised with a number of slums like Bwaise, Mulago, Kamwokya, and Wandengeya (Fig. 3.1). Slums are informal, or even illegal, densely populated areas without access to safe drinking water and improved sanitation. Furthermore, the catchment is characterized by cropland and scattered trees and most residents practice subsistence agriculture and rearing of domestic animals (Fig. 3.2). The Northwest part of the catchment is less urbanized and is covered with only shrubs, trees and croplands. Soils are predominantly composed of sandy clay loams (iron rich laterite). They are usually fertile and application of fertilizers is not common.

3.2.2. Geology and hydrogeology

The area, like many parts of SSA, is underlain by Precambrian basement rocks consisting of predominantly undifferentiated granite-gneiss rocks (silicate rocks) of the Buganda-Toro Cover Formation (Fig. 3.2), which have been deeply weathered to lateritic regolith soils (about 30m thick) (Key, 1992; Taylor and Howard, 1996, 1999a). X-ray diffraction mineralogical analyses show that the weathered regolith is dominated by kaolinite and quartz minerals, with minor amounts of crystalline iron oxide (Flynn et al., 2012; Taylor and Howard, 1999b). These minerals are known to have a poor pH buffering capacity (Taylor and Howard, 1999b). The regolith aquifer is highly productive with a mean hydraulic conductivity ranging from 0.3 - 3 m/d. It is therefore an important aquifer containing shallow groundwater flow systems that discharge into valley springs (Barrett et al., 1999a; Flynn et al., 2012; Taylor and Howard, 1996, 1998). The aquifer is easily impacted by point source pollution because travel times are relatively short owing to the generally high permeability of the aquifer and the hilly topography, which helps to drive gravity-dominated groundwater flow. In and near wetlands surrounding the lower parts of the drainage system, alluvial sands and clays can be found (KDMP 2002; Kulabako et. al., 2007; Fig. 3.2). This alluvial material consists of finer and thus less impermeable soils, which usually cause the water table to intersect the ground surface giving rise to a number of springs (Biryabareema, 2001).

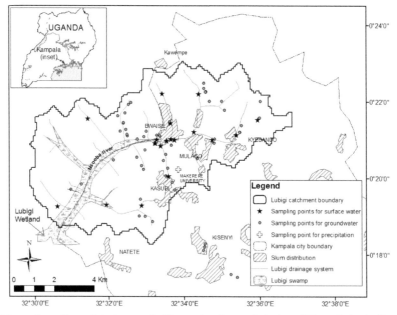

Figure 3.1:   Location of study area in Kampala, the capital city of Uganda including
               locations of sampling points for groundwater, surface water and precipitation.

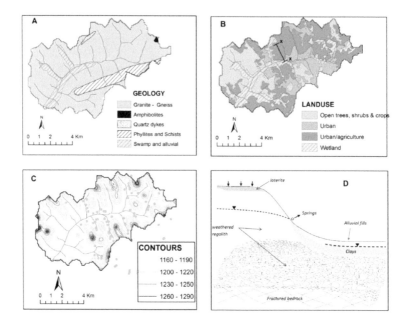

Figure 3.2:   Context of the study area: (a) geology map, (b) land use map (c) topographical
               map and, (d) hydrogeological cross-section x-x illustrating groundwater flow
               and spring formation (adapted from Miret-Gaspa, 2004).

3.2.3. Hydrology

The area experiences two rain seasons (March–May and September–December) with a mean annual rainfall of 1450 mm/y. Evaporation is about 1151 mm/y (see chapter 2; section 2.5). Runoff from the catchment is via a series of intricately linked channels consisting of tertiary drains, secondary drains and the main primary drainage channel (the Nsooba channel) (Fig. 3.1; Fig. 3.3). The tertiary drains are small, open drains (usually unlined) between buildings that convey a combination of storm runoff and wastewater into a system of larger channels, or secondary channels. Secondary channels are natural streams (about 1–2 m wide) located in valleys between gentle sloping hills, which characterize the catchment and Kampala city. These streams are partly channelized and lined with stones at the sides especially in the downstream areas where there is high concentration of human settlement and industrial development. The secondary channels finally discharge into the primary channel, the Nsooba channel (2 – 4 m wide) and eventually in Lubigi swamp located downstream (Fig. 3.1; Fig. 3.3c). The Nsooba channel, which is also partly lined in the downstream areas, accumulates all water from the tributaries (tertiary and secondary drains). Most runoff from the upper and more urbanized part of the catchment (28 km^2) discharges through Bwaise slum, thereby frequently flooding the lower parts of Bwaise slum during the wet seasons. Springs generally form upper reaches of headwaters in secondary channels (personal observation). Springs have a relatively constant discharge and are mainly recharged by precipitation via fractured rocks and regolith aquifers (Flynn et al., 2012; Howard and Karundu, 1992). Aquifer recharge is generally uncertain for most areas in Uganda. However, a few local studies that have applied catchment-based models on the basis of soil moisture balance techniques in central and western Uganda indicate that recharge ranges from 100 – 250 mm/y (Mileham et al., 2008; Nyenje and Batelaan, 2009; Taylor and Howard, 1996).

3.2.4. Sanitation status and water quality

Most residents in the study area have access to piped water supply either through stand pipes or private connections (Katukiza et al., 2010; Kulabako et al., 2010). However, a large number of the low-income population (up to 90%) still rely on protected springs as the preferred source of domestic water supply (ARGOSS, 2002; Barrett et al., 1999a; Haruna et al., 2005; Kulabako et al., 2010). This is partly because springs are free, easily accessible (<200 m), maintain flow throughout the year and are generally regarded to be free of contamination (Flynn et al., 2012). However, majority of springs in Kampala city are contaminated with nitrate and faecal coliforms causing potential health hazard to consumers (ARGOSS, 2002). This contamination is attributed to poor sanitation. In Kampala city, for example, sewer coverage is very low (about 5%; NWSC, 2008), and the existing on-site sanitation is poor. About 80% of the population use pit latrines for excreta disposal and only 4% use septic tanks (KDMP, 2002). Hence, much of the sewage is discharged by infiltrating wastewater to the subsurface, hence impacting on groundwater quality. Most springs in the catchment and Kampala city are therefore polluted with high concentrations of chloride (10– 60 mg/l), nitrate (up to 26.4 mg/l) and total coliforms (3 – 80 000 CFU/100 ml) (Barrett et al., 1999a; ARGOSS, 2002; Haruna et al., 2005; Flynn et al., 2012). In many low lying areas, pit latrines are elevated and their content is sometimes emptied into nearby tertiary channels during flooding events (CIDI, 2006; Matagi, 2002). Hence, part of the wastewater (e.g. from bathrooms and toilets) is discharged untreated into the system of drainage channels giving rise to water pollution. The Nsooba channel is one of the most polluted drainage channels in Kampala and it carries high loads of silt, organic matter and nutrients (Katukiza et al., 2010a; KDMP, 2002; Natumanya et al., 2010). As a consequence, concerns have been raised over the increasing eutrophication and deterioration of surface water due to domestic wastewater discharged from poorly sanitized urban areas (KDMP, 2002; NWSC, 2008).

Figure 3.3.   Photos of drainage channels (a) An unlined secondary channel in the upstream
areas of the catchment, (b) a lined secondary channel in downstream areas of
the catchment and (c) a lined primary Nsooba channel immediately before
discharging into Lubigi wetland.

3.3. Methodology

3.3.1. Sampling and water quality analysis

Sampling involved groundwater, surface water (runoff and wastewater) and rain (precipitation) between the periods September 2009 and August 2012. Groundwater samples were collected from 56 protected springs, surface water samples were collected from 24 sites along primary, secondary and tertiary channels while precipitation samples (13 in total) were collected near Makerere University, College of Engineering, Design, Art and Technology (Fig. 3.1). The precipitation sampler consisted of a one-litre bottle with a funnel of known area designed to collect total atmospheric deposition (during wet and dry periods). Surface water samples were collected during low flow conditions (at least 5 days without rainfall). Groundwater and surface water samples were collected in clean 250 ml PE bottles, and pH, electrical conductivity (EC), temperature, alkalinity (HCO_3^-), orthophosphate (PO_4^{3-}), ammonium (NH_4^+) and nitrate (NO_3^-) were determined on-site with portable meters (WTW 3310, Germany) and a field-kit consisting of a DR/890 portable meter and an Alkalinity test kit. Alkalinity was determined by titrating with 0.2M sulphuric acid. A Hach 890 Colorimeter was used to determine PO_4^{3-}, NH_4^+ and NO_3^- on-site using standard Hach protocols. NO_3^- was determined using cadmium reduction method, NH_4^+ using Nessler's method and PO_4^{3-} using the Ascorbic acid method. Samples for cations (K^+, Mg^{2+}, Ca^{2+}, Mg^{2+}, Fe^{2+} and Mn^{2+}) and anions (Cl^- and SO_4^{2-}) were filtered in the field in 25 ml scintillation vials using a 0.45 µm cellulose acetate filter (Whatman), and kept cool at 4°C. Cation samples were preserved by adding two drops of concentrated nitric acid. The samples were shipped to UNESCO-IHE analytical laboratory, Delft, The Netherlands, for hydro-chemical analyses using an Inductively Coupled Plasma spectrophotometer (ICP - Perkin Elmer Optima 3000) for cations and by Ion Chromatography (IC - Dionex ICS-1000) for anions. To ensure quality of the data, field analyses (i.e. PO_4, NO_3^- and NH_4^+) were checked with Hach standard solutions before every sampling campaign. Moreover, determination of these parameters on-site ensured that the data was not compromised by inadequate preservation. For cations and anions, the quality of data was evaluated by including a blank and standards after every 10 measurements, and where necessary, the analyses were repeated. Data quality controls were also carried out on the full dataset, where all samples with charge balance errors larger than ± 10 % were excluded from the database. An exception was made for precipitation samples because they had low salinity levels (< 64 µS/cm) and they were therefore prone to high analytical errors as noted by Andrade and Stigter (2011).

3.3.2. Data analysis

Hydrochemical data was initially interpreted using descriptive statistics and bi-variate plots between the most pertinent species to gain insight into the state of water quality of the samples and the likely processes controlling water chemistry and nutrients. Bi-variate plots were prepared between Cl^- (as an urban pollution indicator) and various cations and nutrients. Piper and Stiff diagrams were also prepared to provide a concise picture of the existing water types and differences between groundwater and surface water composition. To further aid the interpretation of processes likely affecting nutrients, saturation indices (SIs) of minerals known to control orthophosphates (PO_4^{3-}) in water were calculated using the PHREEQC code (Parkhurst and Appelo, 1999). The most common phosphate mineral phases are those of Fe and Ca (Reddy et al., 1999). Mn phosphate is also reported to be stable and thus capable of regulating P (Boyle and Lindsay, 1986). Hence, the SIs of several Fe, Ca and Mn phosphate minerals were considered. They included hydroxyapatite ($Ca_5(PO_4)_3(OH)$), vivianite ($Fe_3(PO_4)_2.8H2O$), manganese(II)hydrogenphosphate ($MnHPO_4$) and strengite

(FePO$_4$.2H$_2$O). Metal oxides, especially of Fe, are of more importance in co-precipitating or adsorbing P (Reddy et al., 1999). However, under anoxic conditions (insufficient oxygen), reductive dissolution of Fe can take place resulting in release of P in solution. Mn oxides and rhodochrosite are also reported to significantly control the solubility of Mn under varying redox conditions (Boyle and Lindsay, 1986) while the solubility of calcite is an important regulator of Ca^{2+} and P concentrations in water (Reddy et al., 1999). Hence, the saturation indices of hematite (Fe$_2$O$_3$), magnetite (Fe$_3$O$_4$), siderite (FeCO$_3$), rhodochrosite (MnCO$_3$), calcite (CaCO$_3$), manganite (MnOOH) and Hausmanite (Mn$_3$O$_4$) were also calculated to indicate if conditions were optimal for dissolution of Ca^{2+}, Mn^{2+} or Fe^{2+}. Fe and Mn oxides were particularly considered due to wide presence of laterite in the study area.

3.3.3. Multivariate statistical analyses

We applied two multivariate statistical analyses using the R program (ver. 2.15.0; http://www.r-project.org) to understand better the variability in groundwater composition and processes likely controlling the fate of nutrients. These analyses included Principal Components Analysis (PCA) to identify relationships between variables and Q-mode Hierarchical Cluster Analysis (HCA) to identify groups of groundwater samples with similar hydrochemical characteristics. PCA was preferred to common factor analysis because it is the most commonly used technique in statistical programs and international literature. The PCA and HCA techniques were applied on the hydrochemical dataset of 56 groundwater samples. These analyses were done for concentrations in mg/L. But the results are presented in mmol/L because these units are easier to work with during hydrochemical computations. Data consisted of 14 hydrochemical parameters (or variables) including EC, pH, Na$^+$, Ca^{2+}, K$^+$, Mg^{2+}, NH$_4^+$, Cl$^-$, SO$_4^{2-}$, NO$_3^-$, HCO$_3^-$, Mn^{2+}, Fe^{2+} and PO$_4^{3-}$. However, the parameters NH$_4^+$, PO$_4^{3-}$ and Fe^{2+} were excluded from the analyses because they were below detection limits for a large number of samples. Such data is not appropriate for many multivariate techniques as noted by Güler et al. (2002). Hence, multivariate statistical analyses were applied on only 11 variables. Before application, data were first screened for homoscedasticity (homogeneity of variance) and normality requirements using Kolmogorov-Smirnov (K-S) normality tests, histograms and Q-Q plots. The distributions of most parameters were highly skewed to the right. Hence, all data were log-transformed so as to correspond to normally distributed data and to remove heteroscedasticity in the data.

PCA reduces the number of variables (the measured hydrochemical parameters) to a small number of principal components (PCs) by linearly combining measurements (based on weights or eigenvectors) made on the original variables. These PCs can be used to identify the most important variables responsible for the dominant underlying natural or anthropogenic processes (Güler et al., 2012). The optimal number of PCs to extract was based on the Kaiser Criterion (Kaiser, 1960), where only components with eigen values greater than 1 were considered. The variable SO$_4^{2-}$ had a low communality value (< 0.6) and had complex structure and was therefore excluded from the PCA analyses. The PCA technique reduced the remaining 10 variables to 2 uncorrelated PCs explaining 76% of the total variance present in the dataset. PCA results were presented in terms of component scores (the individual transformed observations) and loadings (the weight by which each standardized original variable should be multiplied to get the component score in the PCs). A varimax rotation, a form of orthogonal rotation, was applied to the PCs to minimise the contribution of variables that had minimal significance (i.e. low loadings) and maximize the contribution of variables with high loadings.

HCA, on the other hand, was intended to classify groundwater observations so that samples of the resulting subgroups or clusters are similar to each other, but distinct from other groups.

These clusters are linked to each other based on the similarity/dissimilarity characteristics between them. Before HCA analyses, data were first standardized using median and the mean absolute deviation (mad) to ensure that each variable was weighed equally. Using median values minimizes errors created by outliers. To form the clusters, the Ward method was employed using the squared Euclidean distance. The Ward method uses analysis of variance technique (ANNOVA) to establish the linkage distance between the clusters. The method is a powerful grouping technique and has minimal distorting effects (Güler et al., 2002; Menció and Mas-Pla, 2008). The results of the cluster analysis were presented as a dendrogram (a tree-like graph which displays the clusters along the x-axis and the linkage distance between the clusters along the y-axis). The individual clusters were selected by visually examining the dendrogram. The relation between land use and groundwater types was investigated by superimposing groundwater clusters and the principal component scores (obtained from PCA for each PC at every location) on the land use map using ArcGIS software (ver. 9.2; ESRI, Redlands, CA, USA). Principal component scores can provide a measure of the influence of the chemical processes described by each PC at the given location (Cloutier et al., 2008; Güler et al., 2012).

3.4. Results

3.4.1. Description of the hydrochemistry

Table 3.1 shows descriptive statistics of the hydro-chemical data for groundwater, surface water and precipitation in Lubigi catchment. The complete list of all hydrochemical data is presented in Table 3.2. The final dataset included 56 groundwater samples, 24 surface water samples and 13 precipitation samples (Fig. 3.1). The saturation indices (SIs) of the most important phosphate minerals are presented in Fig. 3.4. Most minerals were not reactive (had negative SIs) and therefore only few were presented in the *SI* plots.

Precipitation. Precipitation had low concentrations of most ions. For example, the average values of EC and Cl^- were 30 μS/cm and 0.08 mmol/l, respectively (Table 3.1). It was dominated by Ca-NO_3-Cl water type (Fig. 3.5), indicating Ca type water that had been contaminated. This contamination was mainly from nitrogen deposition in the form of NH_4^+ and NO_3^- (0.08 mmol/l; Table 3.1 and Table 3.2), but mostly NH_4^+. The concentrations of other ions were close to zero. The pH was acidic with a mean of 6.1 (range = 4.3–8.6) and a weighted-mean pH of 5.8. For comparison, Kulabako et al. (2007), who carried out a similar study in the same study area, obtained a mean pH of 4.7 (range = 4.01–5.41) and a mean NO_3^- concentration of 0.03 mmol/l (2 mg/l), which were similar to our results (Table 3.1).

Groundwater or springs. Most springs from groundwater were acidic (pH = 5 ± 0.4). The chemical composition of groundwater varied over a wide range (Table 3.1). EC, for example, ranged from 37–988 μS/cm with an average of 278 μS/cm and a standard deviation of 178 μS/cm. The most important ions were Na^+ (1.2 ± 0.8 mmol/l), NO_3^- (1.01 ± 0.6 mmol/l) and Cl^- (0.7 ± 0.5mmol/l). These ions were also strongly correlated (Fig. 3.4a). PO_4^{3-}, HCO_3^-, NH_4^+ and Fe (II) were usually absent (<0.002 mmol/l). PHREEQC (Parkhurst and Appelo, 1999), calculations indicated that most groundwater samples were undersaturated with respect to the most important phosphate minerals except $MnHPO_4$ ($SI \cong 0$) (Fig. 3.4a). Based on major cations and anions, the Piper plot (Fig. 3.5) showed that most groundwater samples plotted towards the Na+K corner in the cation triangle and towards Cl^- corner in the anion triangle. The major water quality types identified in the catchment were Na-K-NO_3 and Na-K-Cl with a slight tendency to mixed water types (Fig. 3.5, Table 3.1 and Table 3.2).

31

Table 3.1: Descriptive statistics of the hydrochemistry of surface water and groundwater samples collected from the Lubigi catchment. Values of variables are given in mmol/L except for EC (Electrical conductivity, µS/cm), pH (-) and T (temperature, °C).

Parameter	EC	pH	T	Na+	K+	Mg2+	Ca2+	NH4+	Cl-	HCO3-	SO4 2-	NO3-	PO4 3-	Fe2+	Mn2+
Precipitation (n=13)															
Mean	30	6.1	24.2	0.03	0.04	0.01	0.12	0.05	0.08	0.07	0.02	0.03	0.004	0.000	0.001
Std. Dev.	17	1.0	3.4	0.02	0.04	0.01	0.06	0.07	0.04	0.07	0.01	0.02	0.007	0.000	0.000
Minimum	8	4.3	18.7	0.01	0.01	0.00	0.01	0.00	0.06	0.00	0.02	0.00	0.000	0.000	0.000
Maximum	63	8.6	30.7	0.08	0.15	0.02	0.21	0.21	0.16	0.16	0.04	0.08	0.028	0.001	0.001
Groundwater (Springs; n = 56)															
Mean	274	5.0	24.9	1.18	0.19	0.17	0.28	0.01	0.74	0.31	0.07	1.01	0.002	0.001	0.005
Std. Dev.	178	0.4	1.9	0.77	0.18	0.09	0.14	0.02	0.47	0.35	0.08	0.59	0.003	0.002	0.006
Minimum	37	4.4	23.0	0.27	0.04	0.05	0.08	0.00	0.17	0.00	0.01	0.21	0.000	0.000	0.000
Maximum	988	6.5	32.6	3.97	1.04	0.52	0.68	0.12	2.56	1.56	0.36	3.57	0.015	0.012	0.023
Primary drains (n = 7)															
Mean	576	7.3	27.0	1.98	0.61	0.28	0.71	1.31	1.38	3.47	0.12	0.02	0.019	0.014	0.021
Std. Dev.	66	0.1	3.3	0.43	0.05	0.05	0.12	1.03	0.10	0.11	0.04	0.04	0.008	0.009	0.009
Minimum	481	7.3	22.9	1.04	0.51	0.18	0.48	0.60	1.25	3.34	0.08	0.00	0.010	0.000	0.000
Maximum	647	7.5	31.0	2.25	0.69	0.31	0.84	2.84	1.52	3.60	0.22	0.11	0.033	0.023	0.026
Secondary drains (n = 13)															
Mean	394	7.4	25.5	1.60	0.67	0.28	0.73	0.27	1.03	2.95	0.09	0.07	0.012	0.020	0.027
Std. Dev.	214	0.3	1.5	0.72	0.56	0.11	0.28	0.20	0.47	1.16	0.07	0.08	0.011	0.028	0.011
Minimum	49	6.7	22.5	0.48	0.15	0.12	0.27	0.00	0.17	0.80	0.02	0.00	0.000	0.004	0.012
Maximum	766	7.9	27.9	2.56	2.21	0.45	1.13	0.53	1.98	4.36	0.24	0.25	0.032	0.106	0.054
Tertiary drains (n=4)															
Mean	1330	7.6	25.7	3.33	1.95	0.48	0.91	3.40	2.32	8.18	0.28	0.14	0.036	0.001	0.006
Std. Dev.	416	0.4	1.9	2.71	1.06	0.28	0.59	3.54	0.64	1.99	0.19	0.26	0.036	0.001	0.006
Minimum	1006	7.0	23.9	1.04	0.44	0.13	0.24	0.20	1.37	6.22	0.11	0.00	0.005	0.000	0.000
Maximum	1899	8.0	28.3	6.62	2.84	0.72	1.66	8.37	2.78	10.90	0.54	0.53	0.087	0.002	0.014

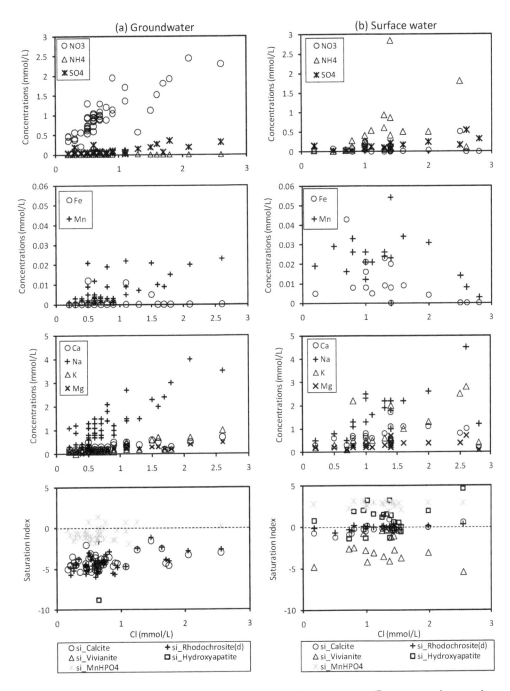

Figure 3.4   Concentrations of major anions and cations versus Cl⁻ concentrations and Saturation indices (SIs) of five selected minerals suspected to influence the occurrence of PO_4^{3-} in both (a) groundwater and (b) surface water. Note that the saturated indices of most minerals were negative and therefore only a few are presented in the SI plots.

Surface water. Surface water in drainage channels was neutral to slightly alkaline (pH = 7.4–7.7). The average EC in primary and secondary drainage channels was 576 and 394 µS/cm, respectively, whereas in the tertiary drains, it was as high as 1330 µS/cm. The dominant cations were Na^+, K^+ and Ca^{2+}. In a number of cases, especially in the primary and tertiary drains, NH_4^+ was high (1.3–2.6 mmol/l). The dominant anion was HCO_3^- (3–4 mmol/l) followed by Cl^- (1 – 2.5 mmol/l). Most surface water samples were characterized by $NaKHCO_3$ and mixed types (Fig. 3.5). Nitrate was usually low or absent, while around 0.02 mmol/l of manganese (Mn^{2+}), and sometimes Fe^{2+} (0.014 - 0.02 mmol/l) were present (Table 3.1 and Table 3.2). This pointed towards reducing conditions in surface water. Compared to groundwater, PO_4^{3-} concentrations in surface water were high with average values of 20–40 µmol/l in primary and tertiary channels and 12 µmol/l in secondary drainage channels. With regard to a number of important phosphate minerals, most surface water samples were saturated with respect to hydroxyapatite [$Ca_5(PO_4)_3(OH)$] and manganese (II) hydrogen phosphate ($MnHPO_4$), but undersaturated with respect to vivianite [$Fe_3(PO_4)_2.8H_2O$] (Fig. 3.4b). It also appeared that the saturation index of rhodochrosite ($MnCO_3$) was in most cases close to zero, suggesting that the solubility of rhodochrosite regulated manganese concentrations in surface water in the Lubigi catchment. Surface water was under-saturated with respect to Fe and Mn oxides indicating these minerals were not reactive and thus did not significantly regulate Mn and Fe concentrations in surface water.

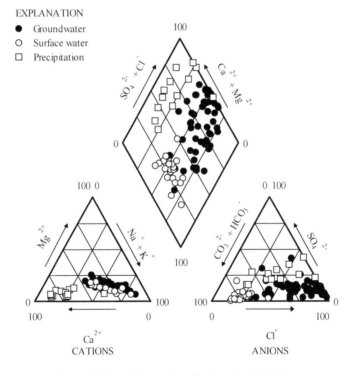

Figure 3.5   Piper plot of hydrochemical data.

Table 3.2: Results of hydrochemistry of surface water and groundwater (springs) in the Lubigi catchment. Values are given mmol/L except for EC (Electrical conductivity, μS/cm), pH (-) and T (temperature, °C).

NAME	Date	EC	pH	T	Na^+	K^+	Mg^{2+}	Ca^{2+}	NH_4^+	Cl^-	HCO_3^-	SO_4^{2-}	NO_3^-	PO_4^{3-}	Fe^{2+}	Mn^{2+}
Precipitation																
R1	24-May-12	21	6.9	24.3	0.02	0.15	0.01	0.05	0	0.14	0	0.02	0	0	0	0.001
R2	26-May-12	17	5.9	26.6	0.02	0.01	0.01	0.04	0	0.06	0.16	0.02	0	0	0	0.001
R3	28-May-12	8	5.8	30.7	0.01	0.01	0.00	0.01	0.02	0.06	0	0.02	0.01	0.005	0	0.001
R4	14-Jun-12	53	6.4	25.0	0.04	0.01	0.02	0.13	0	0.06	0.14	0.04	0.03	0.028	0	0.001
R5	20-Jun-12	31	6.3	23.5	0.04	0.03	0.02	0.15	0	0.06	0.16	0.02	0.06	0.003	0.001	0
R6	25-Jun-12	39	8.6	25.9	0.03	0.02	0.01	0.17	0	0.08	0.10	0.02	0.01	0.002	0	0
R7	25-Jun-12	13	4.3	18.7	0.08	0.02	0.02	0.14	0	0.06		0.02	0.01	0.001	0	0
R8	25-Jun-12	13	5.6	18.7	0.02	0.01	0.01	0.09	0	0.06	0.04	0.02	0.02	0.001	0	0.001
R9	2-Jul-12	40	6.2	22.9	0.03	0.01	0.01	0.15	0.14	0.06	0.06	0.02	0.04	0.001	0	0.001
R10	4-Jul-12	41	6.0	21.7	0.02	0.07	0.01	0.13	0.21	0.10	0.06	0.02	0.06	0.002	0	0.001
R11	10-Jul-12	21	5.7	27.4	0.02	0.06	0.01	0.10	0.14	0.10	0.02	0.02	0.04	0.001	0	0.001
R12	18-Jul-12	63	6.2	23.1	0.05	0.11	0.02	0.21	0.07	0.16	0.16	0.04	0.08	0.001	0	0
R13	24-Jul-12	30	5.7	26.6	0.02	0.02	0.01	0.18	0.07	0.06	0.02	0.02	0.04	0.003	0	0.001
Primary and secondary drains																
BWA2	1-Oct-09	488	7.3	24.2	2.3	0.6	0.3	0.8	ND	1.5	3.5	0.12	0	0.021	0.019	0.024
BWA1	1-Oct-09	647	7.3	29.4	2.1	0.6	0.3	0.7	ND	1.4	3.3	0.12	0	0.026	0.016	0.023
BWA4	1-Oct-09	481	7.4	23.8	2.1	0.7	0.3	0.8	ND	1.3	3.6	0.11	0	0.019	0.023	0.025
B1	20-Oct-09	594	7.3	28.4	2.2	0.6	0.3	0.6	0.86	1.4	3.6	0.10	0.11	0.010	0.008	0.023
PD-UP	25-Sep-09	628	7.5	31.0	1.9	0.6	0.3	0.8	0.60	1.3	3.5	0.08	0.03	0.014	0.009	0.024
B7	28-Jun-10	610	7.4	22.9	1.0	0.5	0.2	0.5	2.84	1.4	3.3	0.22	0.03	0.013	0	0
BWA4(2)	20-Oct-09	587	7.3	29.3	2.2	0.6	0.3	0.8	0.93	1.3	3.4	0.10	0.0	0.033	0.023	0.026
SD17-D	3-Oct-09	103	7.8	23.0	0.5	0.1	0.1	0.3	0.02	0.7	0.8	0.02	0.05	0.003	0.043	0.016
SD3-M	8-Oct-09	380	7.2	25.8	1.3	0.4	0.3	0.8	0.17	1.0	2.6	0.08	0.25	0.005	0.008	0.026

NAME	Date	EC	pH	T	Na^+	K^+	Mg^{2+}	Ca^{2+}	NH_4^+	Cl^-	HCO_3^-	SO_4	NO_3^-	PO_4^{3-}	Fe^{2+}	Mn^{2+}
SD10-D	23-Sep-09	351	6.8	27.9	1.1	2.2	0.2	0.6	0.13	0.8	4.1	0.02	0.07	0	0.008	0.026
SD3-up	8-Oct-09	551	7.3	25.0	1.8	0.6	0.4	1.1	0.41	1.4	4.2	0.06	0.01	0.012	0.020	0.054
SD9-D	26-Sep-09	481	7.5	26.2	1.6	0.5	0.2	0.6	0.53	1.1	2.9	0.12	0.0	0.022	0.005	0.021
BWA6	1-Oct-09	431	7.4	26.0	2.1	0.7	0.3	0.9	ND	1.4	3.3	0.11	0.13	0.002	0.009	0.026
SD7-up	29-Sep-09	504	7.9	27.5	2.3	0.5	0.2	0.7	0.26	1.0	3.4	0.10	0.04	0.019	0.016	0.012
SD6-D	30-Sep-09	682	7.4	25.4	2.2	1.0	0.4	1.1	0.50	1.6	4.4	0.17	0.01	0.003	0.009	0.034
sd13-up	3-Oct-09	150	7.8	25.0	0.5	0.2	0.2	0.3	0.02	0.2	1.1	0.15	0.02	0.011	0.005	0.019
SD6-up	30-Sep-09	766	7.4	26.9	2.6	1.3	0.4	1.1	0.50	2.0	4.1	0.24	0.04	0.022	0.004	0.031
SD10-UP	23-Sep-09	271	6.7	22.5	0.8	0.3	0.2	0.6	0	0.5	2.0	0.02	0.08	0	0.106	0.029
SD7-D	29-Sep-09	49.1	7.5	25.1	2.5	0.5	0.2	0.6	0.40	1.0	3.0	0.07	0.23	0.032	0.021	0.021
SD9-UP(2)	25-Sep-09	408	7.5	25.2	1.3	0.5	0.3	0.7	0.27	0.8	2.6	0.02	0	0.021	0.008	0.033
Tertiary drains																
BWA8	1-Oct-09	1006	8.0	23.9	1.0	2.0	0.7	1.7	0.20	1.4	7.4	0.11	0	0.005	0	0
B4	26-May-10	1899	7.8	28.3	6.6	2.5	0.4	0.8	1.8	2.5	10.90	0.16	0.5	0.087	0.002	0.014
B4	28-Jun-10	1380	7.8	25.2	4.5	2.8	0.7	1.0	3.2	2.6	8.20	0.54	0.0	0.024	0.001	0.008
B4	29-Jul-10	1033	7.0	25.3	1.2	0.4	0.1	0.2	8.4	2.8	6.22	0.32	0	0.025	0.002	0.003
Springs																
K38	14-Jul-10	315	5.2	23.8	0.8	0.2	0.2	0.3	0	0.7	0.3	0.03	1.29	0	0	0.001
K24	12-Jul-10	306	5.1	23.9	1.3	0.2	0.2	0.2	0	0.6	1.0	0.03	0.94	0	0	0.001
K11	9-Jul-10	528.3	4.9	23.6	2.4	0.2	0.2	0.2	0	1.7	0.3	0.06	1.81	0	0	0.009
K15	9-Jul-10	163.2	5.2	23.5	0.9	0.1	0.1	0.1	0	0.3	0.5	0.19	0.41	0.001	0	0.003
K49	16-Jul-10	206	4.8	23.5	0.8	0.1	0.1	0.1	0	0.5	0.2	0.07	0.66	0	0	0.001
K75	19-Jul-10	231	4.8	23.9	0.7	0.1	0.1	0.2	0	0.6	0.0	0.03	1.05	0	0	0.012
K22	12-Jul-10	285	5.1	24.3	1.1	0.1	0.1	0.1	0	0.6	0.3	0.06	0.72	0	0	0.004
K08	8-Jul-10	196	4.8	23.5	0.8	0.1	0.1	0.1	0	0.5	0.1	0.05	0.69	0	0	0.002
K10	9-Jul-10	553.8	4.4	23.8	3.0	0.2	0.2	0.3	0	1.8	0.3	0.36	1.92	0	0	0.015
K33	13-Jul-10	65.7	5.0	24.4	1.8	0.2	0.1	0.2	0	0.8	0.8	0.03	1.20	0.015	0	0.009

NAME	Date	EC	pH	T	Na^+	K^+	Mg^{2+}	Ca^{2+}	NH_4^+	Cl^-	HCO_3^-	SO_4^{2-}	NO_3^-	PO_4^{3-}	Fe^{2+}	Mn^{2+}
K66	19-Jul-10	750	5.7	24.6	2.0	0.7	0.3	0.5	0	1.6	1.2	0.27	1.52	0	0	0.010
K21	12-Jul-10	254	4.7	23.7	1.3	0.1	0.1	0.2	0	0.6	0.3	0.04	0.95	0.003	0	0.004
SPR5	29-Sep-09	311	5.3	24.7	1.4	0.3	0.2	0.4	0.04	1.1	0.0	0.06	1.71	0.014	0	0.005
SPR3	25-Sep-09	224	5.4	23.9	1.0	0.2	0.2	0.3	0	0.7	0.1	0.05	1.29	0.004	0	0.003
K01	8-Jul-10	238	4.9	23.0	1.1	0.1	0.1	0.2	0	0.6	0.0	0.04	1.02	0	0	0.003
K27	13-Jul-10	258	4.9	24.0	1.3	0.1	0.1	0.2	0	0.7	0.2	0.02	0.98	0	0	0.003
K36	14-Jul-10	139.1	5.2	23.6	1.1	0.1	0.1	0.2	0	0.2	1.0	0.02	0.35	0.001	0	0
K69	19-Jul-10	609	6.5	25.1	2.3	0.6	0.3	0.6	0	1.5	1.6	0.19	1.12	0	0.005	0.021
K35	14-Jul-10	232	4.8	23.7	1.5	0.1	0.1	0.1	0	0.7	0.3	0.06	0.91	0.002	0	0.004
K77	19-Jul-10	198.5	5.1	23.1	0.8	0.1	0.2	0.3	0	0.5	0.6	0.04	0.72	0.001	0	0.002
BWA-SPR8	1-Oct-09	466	4.8	25.0	2.7	0.5	0.4	0.5	0.12	1.1	0.0	0.05	3.57	0	0	0.022
SPR4	26-Sep-09	230	5.2	25.3	1.0	0.2	0.2	0.3	0.07	0.7	0.0	0.02	1.29	0.006	0	0.002
K26	13-Jul-10	341	4.9	24.4	1.4	0.1	0.2	0.2	0	0.8	0.4	0.01	0.88	0	0	0.003
K31	13-Jul-10	37.2	5.3	24.0	1.3	0.1	0.1	0.1	0	0.5	0.6	0.04	0.54	0.007	0	0.009
K16	12-Jul-10	308	5.1	24.6	2.1	0.2	0.2	0.3	0	0.8	0.8	0.06	1.20	0	0.001	0.019
K67	19-Jul-10	988	5.3	24.2	3.5	1.0	0.5	0.7	0	2.6	0.8	0.32	2.30	0.001	0	0.023
SPR9	8-Oct-09	303	6.0	23.8	1.5	0.2	0.2	0.4	0.01	1.3	0.6	0.15	0.57	0.002	0	0.007
SPR7	30-Sep-09	314	4.9	23.9	1.5	0.2	0.2	0.3	0.08	1.1	0.0	0.02	1.36	0.002	0.011	0.009
K19	12-Jul-10	275.1	5.0	23.6	1.5	0.1	0.2	0.2	0	0.6	0.1	0.25	0.84	0	0	0.002
K29	13-Jul-10	761	5.3	24.3	4.0	0.7	0.4	0.5	0	2.1	0.6	0.19	2.44	0	0	0.020
K55	19-Jul-10	266	5.5	24.5	1.3	0.2	0.3	0.5	0	0.5	0.6	0.06	1.30	0	0	0.002
K48	16-Jul-10	157.8	5.0	23.7	0.3	0.2	0.2	0.3	0	0.4	0.2	0.03	0.56	0.002	0	0.002
K04	8-Jul-10	148	4.9	25.0	0.9	0.0	0.1	0.1	0	0.3	0.3	0.03	0.38	0.005	0	0.001
K06	8-Jul-10	159	5.0	23.0	1.2	0.2	0.1	0.1	0	0.3	0.3	0.08	0.56	0.002	0	0.001
S21	20-Jun-12	74	4.78	23	0.3	0.1	0.1	0.1	0	0.2	0.1	0.02	0.46	0.000	0.000	0.000
S8	13-Jun-12	343	4.78	32.6	1.2	0.2	0.3	0.5	0	0.9	0.1	0.05	1.95	0.003	0.000	0.002
S6	13-Jun-12	237	5.22	26	0.8	0.2	0.2	0.3	0	0.7	0.3	0.09	0.88	0.002	0.000	0.002

NAME	Date	EC	pH	T	Na^+	K^+	Mg^{2+}	Ca^{2+}	NH_4^+	Cl^-	HCO_3^-	SO_4^{2-}	NO_3^-	PO_4^{3-}	Fe^{2+}	Mn^{2+}
S17	19-Jun-12	202	4.47	25.9	0.7	0.1	0.1	0.3	0	0.6	0.1	0.02	0.94	0.003	0.000	0.003
S13	19-Jun-12	211	4.52	25.8	0.7	0.1	0.2	0.3	0	0.7	0.0	0.05	1.09	0.003	0.000	0.004
S19	20-Jun-12	133.2	4.78	25.7	0.4	0.1	0.1	0.2	0	0.5	0.1	0.03	0.61	0.003	0.000	0.000
S22	20-Jun-12	176.2	4.53	23.5	0.6	0.1	0.1	0.2	0	0.5	0.0	0.02	0.95	0.003	0.000	0.002
S16	19-Jun-12	208	4.39	27.9	0.7	0.2	0.1	0.3	0	0.7	0.0	0.03	1.01	0.001	0.000	0.002
S7	13-Jun-12	288	4.91	26	1.0	0.2	0.2	0.4	0	0.9	0.2	0.06	1.04	0.004	0.001	0.003
S18	20-Jun-12	187.4	4.49	25.6	0.7	0.1	0.1	0.3	0	0.6	0.0	0.02	0.86	0.004	0.000	0.002
S2	12-Jun-12	360	4.92	24.8	0.8	0.1	0.1	0.2	0	0.6	0.0	0.05	0.88	0.004	0.000	0.005
S11	19-Jun-12	169.2	4.72	24.8	0.6	0.1	0.1	0.2	0	0.5	0.1	0.03	0.72	0.004	0.000	0.001
S12	19-Jun-12	283	4.73	25.2	0.8	0.1	0.3	0.5	0	0.9	0.1	0.10	1.14	0.002	0.000	0.001
S14	19-Jun-12	180.7	4.35	25.8	0.7	0.2	0.1	0.2	0	0.5	0.0	0.05	0.73	0.002	0.000	0.002
S3	12-Jun-12	213	5.01	27.1	0.8	0.1	0.1	0.3	0	0.5	0.0	0.03	0.92	0.003	0.000	0.005
S25	3-Jul-12	93.8	5.33	29	0.3	0.1	0.1	0.2	0	0.2	0.0	0.02	0.47	0.003	0.000	0.000
S15	19-Jun-12	159	4.49	23.8	0.6	0.1	0.1	0.2	0	0.5	0.0	0.03	0.68	0.003	0.000	0.001
S1	12-Jun-12	396	5.54	25.1	0.9	0.1	0.1	0.3	0.07	0.6	0.4	0.09	0.59	0.004	0.002	0.005
S5	13-Jun-12	111.3	5.41	26.4	0.4	0.1	0.1	0.2	0	0.3	0.1	0.02	0.41	0.002	0.000	0.000
S20	20-Jun-12	103.9	4.84	23.8	0.3	0.1	0.1	0.2	0	0.2	0.2	0.04	0.35	0.002	0.000	0.001
S24	3-Jul-12	105.7	4.47	25.6	0.4	0.1	0.0	0.2	0	0.4	0.0	0.05	0.21	0.003	0.001	0.003
S9	13-Jun-12	310	6.1	32	1.1	0.3	0.2	0.5	0	0.5	0.9	0.03	0.74	0.003	0.012	0.021

3.4.2. Principal components analysis

Principal component analysis (PCA) reduces observations into principal components (PCs), through which relationships between hydrochemical variables can be identified. These components can be used to understand the dominant mechanisms controlling groundwater chemistry (Salifu et al., 2012). The PCA technique was applied to 10 variables (EC, pH, Na^+, K^+, Mg^{2+}, Ca^{2+}, Cl^-, HCO_3^-, NO_3^- and Mn^{2+}) of groundwater samples, which produced two uncorrelated ($p < 0.001$) components with eigen values greater than 1, all together accounting for 76% of the total variance in the hydrochemical data (Table 3.3).

Table 3.3: Principal component loadings (-) and explained variance for two components with Varimax normalized rotation.

Variables	Principal component	
	PC1	PC2
EC	**0.844**	-0.096
pH	0.391	**0.768**
Na^+	**0.823**	0.152
K^+	**0.889**	-0.072
Mg^{2+}	**0.921**	-0.050
Ca^{2+}	**0.772**	-0.127
Cl^-	**0.918**	-0.166
HCO_3^-	0.280	**0.861**
NO_3^-	**0.838**	-0.387
Mn^{2+}	**-0.791**	-0.127
Eigen value	6.028	1.580
Explained variance (%)	60.275	15.799
Cumulative % of variance	60.275	76.074

The bold and underlined figures are loadings that are significant (> 0.5)

The first PC explained up to 60% of the variance and was characterized by high PCA loadings of EC, Na^+, K^+, Ca^{2+}, Cl^-, NO_3^- and Mn^{2+}. The dominance of Na^+, NO_3^- and Cl^- was indicative of anthropogenic pollution of groundwater and surface water, whereas the presence of Mn^{2+} and Ca^{2+} indicated that these ions also played important roles in this type of water. Pollution could be attributed to infiltration of wastewater from poor on-site sanitation activities in the unsewered catchment. The second PC that cumulatively explained 76% of the total variance was characterized by high positive loadings of HCO_3^- and pH. This component was indicative of the acidifying processes on groundwater, which was probably the cause of acid groundwater as noted in section 3.4.1. The most common acid generation processes in groundwater include acid rain recharge, nitrification following wastewater infiltration and oxidation of organic matter or sulphide minerals such as pyrite (FeS_2) (Appelo and Postma, 2007). PCA results therefore suggested that anthropogenic pollution from wastewater infiltration and groundwater acidification were the most important processes affecting groundwater chemistry in the catchment.

The principal component scores computed for each component at every location are summarized in Fig. 3.6. These scores represent the influence of the components on the groundwater samples. The horizontal axis plots PC1 scores that accounted for most variance in the dataset and represent 'anthropogenic pollution' as described earlier. The PC1 plots show that there were few samples that were heavily polluted (high PC1 scores), a few that were unpolluted (low PC1 scores) and several samples that lied between heavily polluted and unpolluted samples. The vertical index plots PC2 scores, which represented 'acid neutralization' in an acidic environment (pH < 5.4). PC2 plots show that heavily polluted springs (with high PC2 scores) were slightly more neutralized/buffered than unpolluted samples (low PC1 scores). The PC scores are further discussed in section 3.4.3 (Hierarchical cluster analysis of spring water samples).

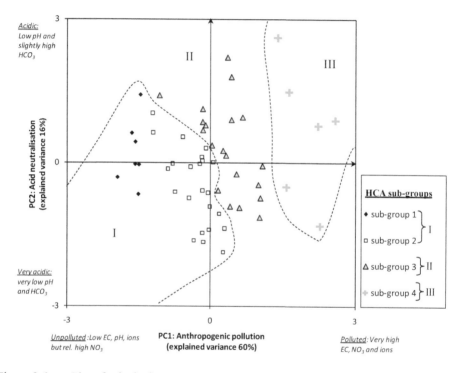

Figure 3.6:   Plot of principal component scores (PC1 and PC2) of the PCA components for groundwater samples labeled with cluster numbers from HCA analysis.

3.4.3. Hierarchical cluster analysis of spring water samples

An HCA was applied to a total of 56 springs using Ward's method to classify groundwater into distinct hydrochemical groups. Eleven transformed variables (pH, EC, Na^+, K, Ca^{2+}, Mg^{2+}, HCO_3^-, Cl^-, SO_4^{2-}, NO_3^- and Mn^{2+}) were used during the classifications. The dendrogram obtained (Fig. 3.7) had three distinct hydrochemical groups: I, II and III. Within these three groups, four subgroups were distinguished at a linkage distance of approximately 18 as indicated by the dashed horizontal line or a phenon line (Fig. 3.7). These subgroups included subgroup 1 and subgroup 2 (under group 1), subgroup 2 (under group II) and subgroup 4 (under group III). Subgroup 4 springs were linked to other groups at a higher distance (linkage distance of about 53; Fig. 3.7) indicating that they were chemically distinct from samples in other groups. Likewise, subgroup 3 was linked to subgroup 1 and 2 at a higher linkage distance (of about 30) indicating that the chemical compositions of subgroup 1 and 2 springs were the lowest of all the groups. The average hydrochemical compositions of the springs in each subgroup are presented in Table 3.4.

Group I springs had subgroups 1 and 2, which were characterized by low EC values (< 216 µS/cm), low pH (4.8–5), low Cl^- concentrations (<0.6 mmol/l or 21mg/l), but moderately high NO_3^- concentrations of 0.37– 0.81mmol/l. Most cations were low (< 0.9 mmol/l) with the most important cations being Na^+ and Ca^{2+}. Subgroup 1 samples, however, had lower values of EC (111 µS/cm), pH (4.8) and concentrations of ions (<0.5 mmol/l) compared to subgroup 2 samples.

Group II springs had one subgroup: 3. It was generally characterized by higher values of EC (275 µS/cm) and pH (5.2). It had higher concentrations of Cl^- (0.8 or 28 mg/l), NO_3^- (1.13 mmol/l) and cations (>1 mmol/l).

Group III springs had the worst hydrochemical quality with very high values of EC (average 688 µS/cm), elevated concentrations of Cl^- (average 1.8 mmol/l), NO_3^- (average 2.2 mmol/l), and Na^+ (average 2.9mmol/l), whereas K^+ and sometimes Ca^{2+} and Mg^{2+} were also elevated (average of up to 0.6 mmol/l). The pH was slightly higher that of group I springs (>5).

In all the groups, PO_4^{3-} concentrations were generally low (<3 µmol/l). Highest concentrations were measured in group I and II samples (2–3 µmol/l), whereas none was detected in Group III springs.

Table 3.4: Mean values of hydrochemistry of groundwater sub-groups. Values of variables are given in mmol/L except for EC (Electrical conductivity, µS/cm) and pH (-).

Group	Sub-group	n*	EC	pH	Na^+	K^+	Ca^{2+}	Mg^{2+}	NH_4^+	Cl^-	SO_4^{2-}	HCO_3^-	NO_3^-	PO_4^{3-}	Fe^{2+}	Mn^{2+}
I	1	7	111	5.0	0.52	0.06	0.17	0.07	0	0.25	0.03	0.25	0.37	0.002	0	0.001
	2	23	216	4.8	0.89	0.12	0.21	0.12	0	0.55	0.05	0.14	0.81	0.002	0	0.003
II	3	20	275	5.2	1.23	0.19	0.32	0.19	0.013	0.80	0.06	0.38	1.13	0.003	0.001	0.006
III	4	6	688	5.3	2.91	0.61	0.53	0.36	0.020	1.77	0.23	0.74	2.15	0	0.001	0.019

*Spring samples contained in each sub-group are shown in the dendrogram in Fig. 3.6

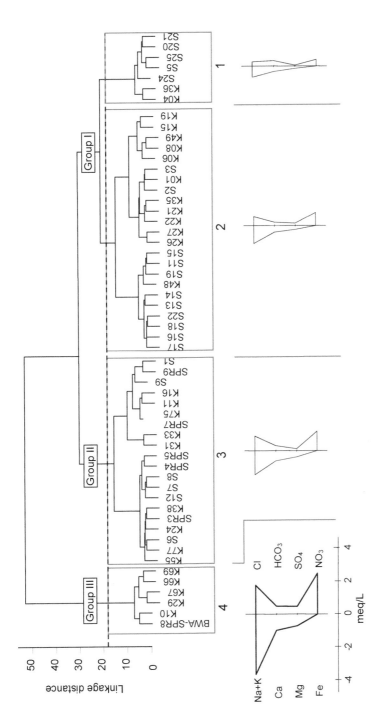

Figure 3.7   Dendrogram based on Q-mode hierarchical cluster analysis (Ward's method) for springs and the mean concentration of Stiff diagram for each cluster.

The Piper diagram in Fig. 3.8 further helps to distinguish the chemistry of the different groundwater groups. Majority of Group II (subgroup 3) and Group III (subgroup 4) springs plotted in the middle of the diamond-shaped diagram indicating that there was no dominant water type. In these groups, the anions seemed to be dominated by Cl^- and SO_4^{2-} (and NO_3^-, Table 3.4), both indicators of anthropogenic pollution. Group I samples were either dominated by Ca + Mg + Na +K for subgroup 1 springs and by Na + K + Ca +Mg for subgroup 2 springs. The ions were dominated by Cl^- and SO_4^{2-} (and NO_3^-, Table 3.4).

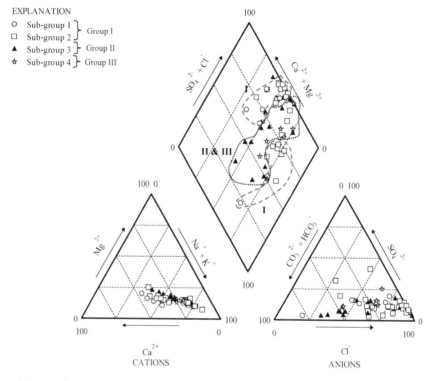

Figure 3.8:   Piper diagram of the four subgroups of the spring samples. The dotted lines represents the two main clusters of springs; group I and group II & III.

3.4.4. Land use control on hydrochemistry

A relevant question is whether these groups and their compositions can be (partly) explained by the land use pattern in Lubigi catchment. To test the relationship between the statistical subgroups obtained with the land use attributes, samples for each group were plotted on the land use map (Fig. 3.9). From this Figure, we observed the following:

- Subgroup 1 samples were located either in wetlands or in uninhabited areas (open trees or shrubs). Subgroup 2 samples were also located in similar areas as subgroup 1 samples, but with a few samples located in less urban/agricultural areas.

- Subgroups 3 were mostly located in urban areas and some near slum areas.

- Subgroup 4 samples were all located within the slum areas of Bwaise, Kasubi and Kisenyi (Fig. 3.9).

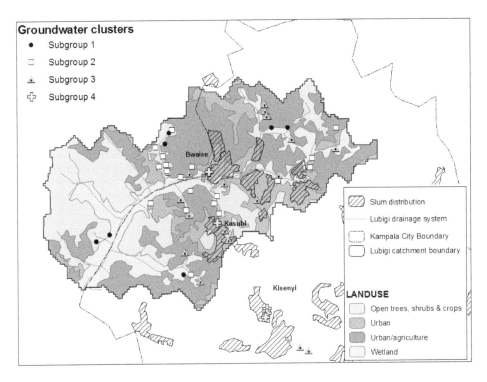

Figure 3.9:   Map showing relationship between land use and sample points for each groundwater cluster .

These findings show that there was a relationship between groundwater chemistry of the subgroups obtained from HCA and land use attributes of the catchment. Fig. 3.6 shows a plot of principal component scores obtained for each PC at each location and labeled by the individual HCA cluster subgroup numbers. Subgroup 4 samples plotted in the upper right quadrant and according to the PC1 axis, they represented groundwater that was heavily polluted with very high EC, NO_3^- and concentrations of ions but slightly buffered compared to other springs. These findings concur with the findings from the spatial distribution of HCA subgroups with land use as described in section 3.4.4 (Land use control on hydrochemistry). Likewise, scores of subgroup 1 samples, which were indicated as less polluted from HCA analysis, plotted in the lower left quadrant in Fig. 3.6, representing acidic groundwater that was less polluted and less buffered (more acidic). Subgroup 2 and 3 samples plotted in the middle of the upper right and lower left quadrants indicating moderate pollution and acidification.

3.5. Discussion

3.5.1. Precipitation

The results from descriptive statistics showed that precipitation was acidic with pH ranging from 4.3 – 8.6 and weighted-mean pH of 5.8. An even lower precipitation pH (average = 4.7 ± 0.7 and range = 4.01–5.41) was obtained by Kulabako et al. (2007) in the same study area, hence, confirming the presence of acidic rains. Literature shows that acid rain is principally caused by sulphurdioxide (SO_2) and nitrogen oxide (nitric oxide, NO and nitrogen dioxide $NO_2 = N_xO$) gases introduced in the atmosphere by both natural and anthropogenic pollution (Appelo and Postma, 2007). SO_2 is mainly from natural sources of pollution (e.g. volcanoes or windblown dust containing gypsum) and burning of coal, whereas N_xO is from human pollution activities (burning of fossil fuels) (Irwin and Williams, 1988). These gases and their aerosols fall on the earth's surface as wet and dry acid deposition in form of dilute nitric and sulphuric acid. In this study, acid rain was mostly dilute nitric acid (contaminated with nitrogen species; 0.08 mmol/l) and it was likely primarily caused by burning of fossil fuels, which we attributed to the fast-growing automotive industry in Kampala ciy. In other parts of Africa, acid rain with high sulphur species has been reported owing to presence of large mining industries containing sulphur particles (e.g. in Highveld, South Africa: Davies and Mundalamo, 2010; Mphepya et al., 2006; Zunckel et al., 2000). Acid rain is known to directly affect the chemistry of groundwater in poorly buffered aquifers (Appelo and Postma, 2007) and in this study, these effects are discussed in the succeeding texts. To our knowledge, few studies have investigated effects of acidic rains on groundwater in SSA.

3.5.2. Impacts on groundwater quality

Groundwater showed a wide variability in water quality but was strongly influenced by anthropogenic inputs due to the dominance of species like Na^+, K^+, NO_3^- and Cl^-. These species are good markers of anthropogenic inputs in urban areas (Barrett et al., 1999b). The results showed that these species including Ca^{2+}, Mg^{2+} and Mn^{2+} also had high PCA loadings explaining up to 60% of the variance in the hydrochemical data. NO_3^- and Na^+ were also strongly correlated with Cl^- (Fig. 3.4 a) indicating a dominant anthropogenic source of both species (Morris et al., 2006). The role of water-rock interactions giving this type of water, especially because of the presence of Na^+, Ca^{2+}, K^+ and Mn^{2+}, was therefore likely limited. Moreover, the mineralogy of the weathered regolith predominantly consisted of quartzite (SiO_2) and kaolinite ($Al_2Si_2O_5 (OH)_4$) (see section 3.2.2), which are usually depleted of these minerals or if present, then their concentrations are low. It is therefore not surprising that PCA analysis identified only anthropogenic pollution and groundwater acidification as the major processes that controlled groundwater chemistry. The presence of high concentrations of Cl^- is also an indicator of groundwater contamination from on-site sanitation systems as noted by related studies (e.g. ARGOSS, 2002; Flynn et al., 2012; Foppen, 2002; Graham and Polizzotto, 2013). These results therefore demonstrated the dominance of anthropogenic pollution on groundwater chemistry and nutrients. This pollution was likely from wastewater infiltrating from on-site sanitation systems like pit latrines and septic tanks because the catchment was unsewered.

A clear distinction of the sources of pollution, however, can be obtained from the results of HCA, where three distinct groundwater quality groups were evident: Group I springs, which contained subgroup 1 and 2 springs that were generally less polluted; Group II springs, which contained subgroup 3 springs that were polluted; and Group III springs, which contained subgroup 4 springs that were heavily polluted (Fig. 3.7; Table 3.4). The spatial distribution of the cluster groups superimposed on the land use map (Fig. 3.9) indeed showed that group III samples, which were heavily polluted, were located in slum areas (or poorly sanitized informal settlements), whereas group II samples, which closely approximated group III springs, were located in urban areas. Subgroup 1 springs were generally classified as unpolluted and indeed, they were located in less impacted areas (wetlands, open trees and shrubs). The relationship between the water quality of the samples and the location of the samples are also illustrated by the PCA results in Fig. 3.6. Here, PCA results showed that the dominant chemical processes controlling Group III and II springs were anthropogenic in nature whereas for group I, they were generally acid and nitrate generation processes. A better relationship can even be seen on the distribution map of the PC1 scores (representing pollution index), which accounted for the greatest variance in the dataset (Fig. 3.10). Elevated scores for PC1 plotted either in slum areas or in urban areas, whereas less elevated scores of PC1 plotted in less urbanized areas.

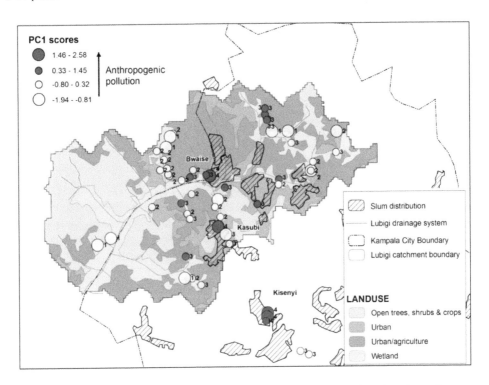

Figure 3.10:   Map showing relationship between land use and sample points for each groundwater cluster.

An explanation of the evolution of groundwater chemistry in the catchment as explained by HCA and PCA analyses is given in the succeeding texts. Results from PCA and HCA analyses showed that subgroup 1 springs were located in less urbanized areas of the catchment (Fig. 3.9; Fig. 3.10) and generally had the best hydrochemical qualities (Table 3.4). In addition to this, these springs had the lowest pH, lowest concentrations of ions, but with abnormally high NO_3^- concentrations (0.37 mmol/l). In the following texts are some hypotheses regarding why these springs were more acidic than the rest and the likely source of nitrate in this rather less impacted groundwater:

1. Subgroup 1 springs were recharged by acid rain, which had underdone evaporation.

2. There was decomposition of organic matter because of the abundant supply of organic matter and the relatively high temperatures. Decomposition of organic matter resulted in the production of carbonic acid, which in combination with the low buffering capacity of the aquifer reduced the pH of groundwater. Groundwater acidification due to pyrite oxidation was not considered because of the absence of pyrite in the catchment (Fig. 3.2).

3. Low buffering capacity of the aquifer because of the deep weathered regolith and the short flow paths of spring water.

4. Minor pollution to springs originated from scattered households and agricultural fields, whereby the low pH and presence of NO_3^- were caused by nitrification of NH_4^+.

In relation to the first hypothesis, subgroup 1 springs had Ca-NO_3-Cl water type, low pH, low EC values and moderately high concentrations of NO_3^- (0.37 mmol/l: Table 3.4; Fig. 3.8). The chemistry of these springs was similar to the chemistry of precipitation that had undergone concentration by evaporation as shown by Cl^- as a conservative tracer. The use of Cl^- to estimate evaporation-related increase in concentration is widely accepted in unpolluted environments given that it is rarely removed by deposition (precipitation) or significantly supplied via interactions with rocks (Barrett et al., 1999b; Séguis et al., 2011; Wood and Sanford, 1995). The results of Cl^- of precipitation (0.08 mmol/l, Table 3.1) and of subgroup 1 springs (0.25 mmol/l; Table 3.4) suggests an evaporation factor of 3.3. We applied this factor to all precipitation concentrations. Hence, the sum of N species ($NH_4^+ + NO_3^-$) originating from precipitation after evaporation, was $0.05 \times 3.3 + 0.03 \times 3.3 = 0.26$ mmol/l after concentration by evaporation, which is close to NO_3^- concentrations of subgroup 1 springs (0.37 mmol/l) (Table 3.4). The same calculation applied to EC. This finding suggested that subgroup 1 springs were probably a result of recharge from precipitation and that the N observed in subgroup 1 springs resulted from evaporation of N in acidic precipitation. The low pH of subgroup 1 could also have been largely explained by recharge from acid precipitation: when subjected to an evaporation factor of 3.3, the [H] in precipitation increases from $10^{(-6.1)}$ to $3.3*10^{(-6.1)}$. The resulting pH would then be 5.6, which was close to the average pH measured in subgroup 2 springs and the rest of the springs. Lower pH values of these springs were, however, likely affected by other processes such as the degradation of organic matter present in the soil (produces carbonic acid) and the conversion of NH_4^+ to NO_3^- by nitrification (produces dilute nitric acid), in combination with the weak buffering capacity of the silicate-dominated soil and the regolith, which both do not buffer hydronium ions (H^+ or H_3O^+) (Taylor and Howard, 1999b) and the short travel times of spring water. Subgroup 2 springs were similar to subgroup 1 springs, but slightly more polluted as indicated by the higher concentrations of NO_3^- (0.8 mmol/l) and Cl^- (0.6 mmol/l), which is expected because some of these springs were located in less urbanized areas.

Superimposed on this catchment wide acidic groundwater from precipitation were Group II and Group III springs, which emanated with higher concentrations of NO_3^- and, of most cations and anions. We think that these types of springs (Group II and Group III) were not only affected by recharge from acidic precipitation, but also by increased infiltration of wastewater from intensively urbanized areas and slums or informal settlements (see Fig. 3.9). Ammonium in wastewater leachate was likely oxidized to nitrate by nitrification, giving rise to oxic groundwater with higher nitrate concentrations (Table 3.4). This also explains why

HCO_3^-, PO_4^{3-}, NH_4^+ and Fe^{2+} were generally low or absent. In oxic environments, Fe^{2+} is oxidised to immobile Fe III, NH_4^+ to NO_3^-, whereas PO_4^{3-} usually sorbs or co-precipitates with immobile Fe^{3+} (Appelo and Postma, 2007). Springs in Group II and Group III were the most polluted and were located within or near slum areas (Fig. 3.9). This interpretation was confirmed by both PCA and HCA analyses, which showed that there was a relationship between these type of springs and the distribution of slums in the catchment. As mentioned earlier, phosphate concentrations were low, probably due to sorption to Fe- and Mn-oxides given that the study area is characterised by lateritic soils. But phosphate transport was also likely retarded owing to precipitation of $MnHPO_4$ because most springs were saturated with respect to this compound. The source of Mn^{2+} in groundwater was likely from acid drainage because of the presence of dilute nitric acid in groundwater or from reduction of manganese minerals (e.g. amphibolites located in the upper part of the catchment Figure 3.2), which occurred when reducing wastewater leachate from on-site sanitation got in contact with these minerals. The presence of reducing conditions in areas impacted by wastewater is supported by the fact that polluted springs (e.g. subgroup 4) had higher concentrations of HCO_3^- and Mn than unpolluted springs (e.g. subgroup 1 springs) (see Table 3.4). In aquifers without primary carbonate minerals as in this study, the likely source of HCO_3^- is CO_2 produced from the oxidation of wastewater. Apparently, denitrification, which was likely present in polluted springs, resulted in production of HCO_3^- during degradation of organic matter in wastewater (Appelo and Postma, 2007). Subsequently, HCO_3^- produced from denitrification led to the slight pH neutralization of subgroup 4 springs (pH of 4.8 – 5.4; Table 3.4). Additionally, the complete absence of PO_4^{3-} in subgroup 4 could be explained by higher reducing conditions in these springs, which resulted in higher Mn^{2+} concentrations and the subsequent removal of P by precipitation of $MnHPO_4$. Based on the discussion earlier, Figure 3.11 shows a systematic representation of the evolution of groundwater/spring water quality in Lubigi catchment.

3.5.3. Surface water

Surface water had neutral to slightly alkaline pH whereas groundwater was largely acidic. NO_3^- was generally absent in surface water whereas NH_4^+ and PO_4^{3-} concentrations were generally high. In groundwater, the reverse was true, hence, reflecting the contrasting redox environments where surface water was anaerobic and groundwater oxic. Denitrification likely took place immediately after the nitrate-rich groundwater (Table 3.1) exfiltrated to surface water where nitrate in groundwater was converted to nitrogen gas. At the same time, groundwater did not contain PO_4^{3-} (see aforementioned texts), and thus did not contribute to the PO_4^{3-} load observed in surface water. Hence, the only possible source of N as ammonium and P as orthophosphate in surface water was through wastewater discharged directly to surface water. Because of poor on-site sanitation systems, it is common practice to discharge wastewater directly to tertiary channels, which eventually discharge into secondary and primary channels. This wastewater is usually a mixture of grey water from bathrooms and kitchen and black water derived from open defecation and faecal sludge emptied from pit latrines (see section 3.2.1). Pollution from households is also reflected by the elevated concentrations of major cations and anions, especially in the tertiary channels (Table 3.1). There was a positive correlation between Cl^- (a pollution indicator) and most of the cations and anions, except Mg^{2+} (Fig. 3.4). Mg^{2+} likely originated from a natural source and was less affected by redox reactions. According to Dodds et al. (1998), streams are eutrophic when total N exceeds 1.5mg/l ($\approx$0.1mmol/l) and total P exceeds 0.075mg/l ($\approx$2 µmol/l). Nitrogen measured as NH_4^+ in surface water was $3 - 26$ times higher than 0.1mmol/l, whereas phosphorus measured as PO_4^{3-} was 5–20 times more than 2 µmol/l (Table 3.1). Hence, surface water in Lubigi catchment was eutrophic and posed a danger to downstream ecosystems. Phosphorus, in particular, is of great concern because it is considered the limiting nutrient for plant growth (Reddy et al., 1999; Rodríguez-Blanco et al., 2013).

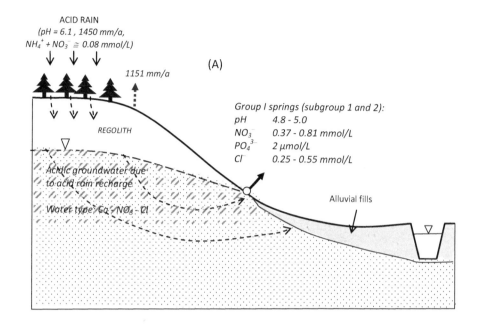

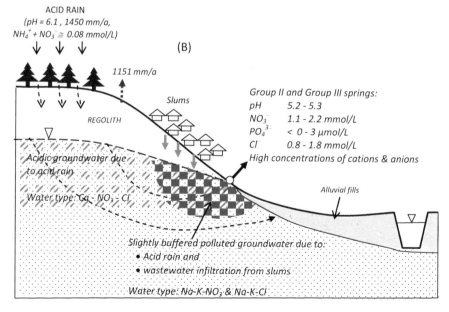

Figure 3.11:   Schematic sketches explaining the evolution of spring water quality in the study area: (A) Less polluted springs mainly affected by acid rain recharge, and (B) polluted springs affected both acid rain and addition of wastewater leachate from urban areas and slum areas.

3.6. Conclusions

The results showed that Lubigi catchment was impacted by acidic rains, which contributed to nitrate and the acidity observed in groundwater. This acidity was poorly buffered because the aquifer consisted of deeply weathered granite and gnesis rocks, and the travel times were also short. Groundwater, therefore, had low pH. The dominant ions in groundwater were Na^+, K^+, NO_3^- and, Cl^-. The principal component loadings of these ions contributed up to 60% of the total variance in the hydrochemical data. These ions were also strongly correlated, indicating that groundwater chemistry was largely influenced by anthropogenic pollution. This pollution was attributed to wastewater infiltration from on-site sanitation in informal settlements and urban areas of the catchment.

To distinguish the sources of pollution, HCA of groundwater was carried out, which showed that groundwater chemistry was explained by three distinct water quality groups:

1. Group I springs, which were located in less urbanized areas of the catchment, were characterized by Ca-Cl-NO_3 water type with low values of pH (4.8 – 5) and EC (111–216 µS/cm), but with relatively high NO_3^- (0.37–0.81 mmol/l) concentrations;

2. Group II springs, which were located in urban and near slum areas, were dominated by a mixture of Na-K-Cl-NO_3 and Ca-Cl-NO_3 water type and had relatively higher pH values (5.2), higher EC values (275 µS/cm) and higher NO_3^- (1.13 mmol/l) concentrations; and

3. Group III springs, which were located in slum areas were similar to group II springs but with much higher values of EC (688 µS/cm), NO_3^- (2.15 mmol/l), pH (5.3) and other cations.

Ion concentrations of Group I springs were similar to that of precipitation after concentration by evaporation. Hence, Group I springs were interpreted to be directly affected by acid precipitation, which was also the source of nitrate and acidity in these springs. Group II and Group III were interpreted to be affected by both acid precipitation and infiltration of wastewater leachate that had undergone nitrification during transport from urban/slum areas to the aquifer.

Generally, groundwater had very low PO_4^{3-} concentrations, which we attributed to sorption to Fe and Mn oxides, and precipitation with Mn^{2+}. Nitrification in groundwater, on the other hand, resulted into nitrate-rich groundwater. The nitrate-rich groundwater, however, did not contribute to nutrients observed in surface water. Instead, surface water was characterized by high concentrations of NH_4^+ and PO_4^{3-} (>20 times the minimum required to cause eutrophication), which we attributed to direct discharge of untreated wastewater from households into tertiary drains. The nitrate in groundwater that was discharging to surface water drains was believed to be lost as nitrogen gas by denitrification because of the slightly reducing conditions in surface water.

The results showed that unsewered urban slum catchments built on deeply weathered regolith material are very vulnerable to pollution, and that there were two main sources of pollution in the catchment: acidic rains with N species and domestic wastewater generated from slums. There is, however, a need to understand better the local processes/mechanisms leading to nutrient pollution from on-site sanitation systems and their fate in both groundwater and surface water. This is possible through micro-scale or experimental-based processes studies, which provide better understanding of the underlying processes (Cash and Moser, 2000).

Chapter 4

Nutrient pollution in shallow aquifers underlying pit latrines and domestic solid waste dumps in urban slums

Abstract

The lack of proper on-site sanitation in unsewered low-income areas is becoming an important source of nutrient-rich wastewater leaching to groundwater and can potentially lead to eutrophication. For typical conditions in sub-Saharan Africa, the nutrient loading of nitrogen (N) and phosphorus (P) from on-site sanitation systems to aquifers is largely unknown. In this study, we assessed the dissolved nutrient loads (nitrate (NO_3^-), ammonium (NH_4^+) and orthophosphate (PO_4^{3-})) and the local processes likely affecting them in aquifers underlying two on-site sanitation systems in an unsewered low-income urban slum area in Kampala, Uganda; a domestic solid waste dump and a site with two pit latrines. The impact of the two types of sites was assessed by comparing the upgradient and downgradient nutrient concentrations and loads along groundwater flow lines. Significant pollution to groundwater originated from the pit latrine site with downgradient nutrient loads increasing by factors of 1.7 for NO_3^-, 10.5 for NH_4^+ and 49 for PO_4^{3-}. No effect of leaching of nutrients to groundwater was found from the waste dump. We estimated that approximately 2-20% of total N and less than 1% of total P mass input was lost to groundwater from the pit latrines. The bulk of N leached to groundwater was in the form of NH_4^+. Mn-reducing conditions prevailed in the shallow aquifer which suggested that nitrification was the main process affecting NH_4^+ concentrations. Phosphorus was likely retained in the soils by precipitating as $MnHPO_4$ and $Ca_5(PO_4)_3(OH)$. Our results indicated that on-site sanitation practices (e.g. pit latrines) in alluvial aquifer systems can be highly effective in the removal of nutrients depending on hydrological, hydrochemical and geochemical conditions in the aquifer receiving wastewater. Improvements to make the current pit latrine systems better for nutrient containment are suggested based on findings from this study.

This chapter is based on:
Nyenje, P.M., Foppen, J.W., Kulabako, R., Muwanga, A. and Uhlenbrook, S. 2013. Nutrient pollution in shallow aquifers underlying pit latrines and domestic solid waste dumps in urban slums: Journal of Environmental Management, 122, 15-24.

4.1.   Introduction

Most wastewater generated in unsewered low-income urban areas is disposed of via on-site sanitation facilities. The risk posed by inadequate on-site sanitation systems to groundwater has thus been the subject of several investigations and reviews (Abiye et al., 2009; ARGOSS, 2002; Barrett et al., 1999a; Cronin et al., 2007; Dzwairo et al., 2006; Foppen, 2002; Kimani-Murage and Ngindu, 2007; Kulabako et al., 2007; Montangero and Belevi, 2007; Taylor et al., 2006). Inadequate on-site sanitation systems are usually associated with groundwater pollution through leaching of wastewater contaminants into underlying aquifers, which contribute to microbiological contamination of water sources and increased nutrient concentrations (nitrogen, N and phosphorus, P) and discharges to downstream ecosystems (Montangero et al., 2007). The latter may contribute to eutrophication of surface waters (Neal et al., 2005; Nyenje et al., 2010). For much of sub-Saharan Africa (SSA), there is still a large knowledge gap regarding the loads and processes controlling the transport of faecal contaminants in groundwater (e.g. Tredoux and Talma, 2006). Failure to address this pollutant source contributes to increasing nutrient pollution of groundwater and the subsequent eutrophication of surface-water bodies. The hydrochemical processes which take place when wastewater leachate enters aquifer systems include redox reactions, sorption, dilution, precipitation and dissolution, and ionic exchange (Mikac et al., 1998). In order to control nutrient discharges in groundwater, it is important to assess nutrient loads and processes affecting these. This research investigated dissolved nutrient loads to groundwater and the hydrochemical processes resulting from the infiltration of wastewater to shallow groundwater beneath two common on-site sanitation systems: a pit latrine site and a domestic solid waste dump in an unsewered urban slum in Kampala, the capital of Uganda. More specifically, our objectives were to: 1) Distinguish between impacts of pit latrines and domestic solid waste dumps on shallow groundwater in unsewered slum areas; 2) Identify the dominant hydrochemical processes affecting dissolved nutrients (nitrate (NO_3^-), ammonium (NH_4^+) and ortho-phosphate (PO_4^{3-})) immediately below a pit latrine and a domestic solid waste dump; and 3) Determine the contribution of pit latrines and domestic solid waste dumps to dissolved nutrient masses leaching into groundwater.

4.2.   Study area (Bwaise III parish slum)

Bwaise III parish (32° 33.5'E, 0° 21'N) in Kampala City, Uganda is an urban slum located in a low-lying area, largely reclaimed from Lubigi swamp (located downstream) (see Fig. 3.1 and Fig. 5.1). The parish has a population of about 15,000 and an area of 57 ha (Kulabako et al., 2007). Our research focused on two sites located in the western part of the slum, locally known as Katoogo Zone (refer to Fig. 5.1). The Kampala area where Bwaise is located is underlain by crystalline rocks and the Buganda-Toro Cover Formation (consisting of mainly quartzites and schists) which have been deeply weathered to form lateritic regolith soils (Taylor and Howard, 1996, 1999a). The weathered regolith (up to 30m thick) and fractured quartzites are excellent storage reservoirs for groundwater that sustain springs within the Buganda-Toro system with transmissivities of up to 10 m^2/d (Flynn et al., 2012; Mukwaya, 2001; Taylor et al., 2009). Groundwater reservoirs in the regolith are mainly located at the base of the saprolite (less weathered regolith) where selective mineral decomposition has produced an unconsolidated and highly permeable gravel-like material (Howard and Karundu, 1992; Jones, 1985). Lowland areas where the Bwaise III slum is located are occupied by alluvial deposits that derive from weathering and erosion from hill summits. These deposits are characterized by a shallow water table (< 1 m) and a succession of fine grained material of sand, silt and clays where by the clayey sand is the aquifer and the clay is the confining layer with an average thickness of around 6 m (Biryabareema, 2001; Kulabako et al., 2008).

Given that Bwaise III slum is largely a reclaimed wetland, top soils essentially consist of fill material of lateritic soil and clayey sand with polyethylene bags, cloths, paper, charcoal and decaying organic matter (Kulabako et al., 2007). These soils are moderately alkaline (pH = 6 – 8) with high calcium and phosphorus contents (>15 meq/100 g and 15 mg/kg respectively). Top soils are also dominated by kaolinite clay minerals characterized by a low cation exchange capacity (11 - 12 meq/100 g; Kulabako et al., 2008) but with a high potential to retard contaminants such as phosphorus, coliforms and viruses (ARGOSS, 2002; Flynn et al., 2012). Previous studies have shown that the permeability of the shallow aquifer in the study area is variable with hydraulic conductivity values ranging from 0.2 to 3.6 m/d (Biryabareema, 2001; Kulabako et al., 2008). Existing rainfall data (2001 - 2010) shows that the area experiences two rain seasons (March - May and September - December) with a mean annual rainfall of about 1450 mm. Like most slums in SSA, Bwaise III is characterized by poor on-site sanitation. Grey water is usually indiscriminately disposed of in open spaces and drainage channels. Open dumping of solid waste has resulted in permanent illegal solid waste dumps in people's backyards. In addition, more than 80% of the pit latrines are of the traditional unimproved type (Katukiza et al., 2010; Kulabako et al., 2004). According to WHO and UNICEF (2012), unimproved pit latrines are defined as those that are unlined, hanging with an open pit, without a slab or shared by more than one household. They are usually unhygienic and pollute groundwater through direct or indirect wastewater infiltration into underlying aquifers (Franceys et al., 1992; Katukiza et al., 2012). In the study area, most latrines are elevated owing to the high water table (<1.5 m below the surface) and are shared by about 6 - 10 households with an average user to pit ratio of approximately 1:30 (Katukiza et al., 2010; Kulabako et al., 2010). Poor vehicular access in the area and the associated high costs of desludging means that most latrine contents (over 60%; CIDI, 2006) are periodically discharged either into adjacent excavated unlined pits or into nearby drains during heavy rains. As a consequence of these unsanitary activities, groundwater is heavily polluted from wastewater leaching into the aquifer.

Although several research studies have been undertaken to understand groundwater pollution from on-site sanitation activities, these have focused mostly on understanding pathogenic bacteria loads and the potential to contaminate spring water (e.g. Howard et al., 2003; Haruna et al., 2005; Nsubuga et al., 2004). A few studies have attempted to focus on other contaminants but have only looked at few parameters like nitrate (Flynn et al., 2011; Kulabako et al., 2007), with even fewer studies attempting to identify hydrochemical processes affecting sanitation-related nutrients in groundwater (e.g. Kulabako et al., 2008).

4.3. Methodology

4.3.1. Study sites and the monitoring network

A domestic solid waste dump site covering about 7 m x 25 m (Fig. 4.1A) and a site with two unlined/elevated pit latrines each with one stance (Fig. 4.1B) were selected for this study. These two gently sloping sites are located in the Katoogo Zone within Bwaise slum (see section 4.2 and Fig. 5.1). The pit latrines (unimproved type) were constructed with brickwork and were un-lined at the bottom (personal discussion with residents). They were raised above the ground because of the high water table, and the need to increase space for faecal sludge storage (Fig. 4.3a). The domestic solid waste dump (herein also referred to as a waste dump), on the other hand, was a result of uncontrolled local solid waste disposal on the land surface. Dumped material was largely organic comprising of waste food, food peeling (bananas, potatoes, pineapples, oranges) and vegetative leaves (Fig. 4.3). This dump site was also a recipient of human excreta through the use of the so called flying toilets (here polyethylene bags are used for excreta disposal and dumped in drainage channels and waste dumps) and through open defecation especially by children (personal observation). In addition to this, the waste dump was also a recipient of pig wastewater from a stall located near downgradient wells SW2 and SW6 (Fig. 4.1A). Previous studies have indicated that about 80% of the solid waste generated in this area is biodegradable (Katukiza et al., 2012; Kulabako et al., 2004). The waste dump was also observed to contain many polyethylene bags, plastic bottles, broken glass, old cloths and paper as also noted by Kulabako et al. (2004).

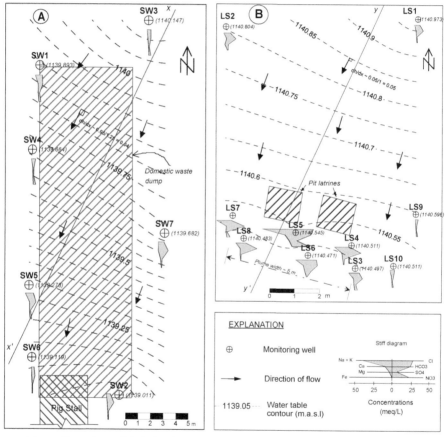

Figure 4.1: Overview of (A) the domestic waste dump site and (B) the pit latrine site (B), including well locations, stiff diagrams of groundwater chemistry and the average water table contours. The shaded box in Fig. 4.1(A) shows the extent of the domestic waste dump site, while the two shaded boxes in Fig. 4.1(B) are the locations of the traditional unimproved pit latrines. The contours were constructed from groundwater levels with a contouring software using Kriging techniques. x - x' and y - y' are lines of geological cross-sections (Fig. 4.4).

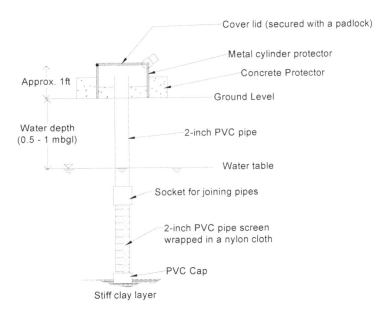

Figure 4.2   A typical monitoring well used for sampling groundwater in the shallow sandy
alluvial aquifer in Bwaise III parish slum.

Figure 4.3:   Installation of the monitoring wells at (a) the pit latrine site and (b) the solid
waste dump site.

Groundwater was sampled from monitoring wells (piezometers) installed around these sites (Fig. 4.1, Fig. 4.2), and data were collected on hydrochemistry, groundwater flow and soil properties to understand the nutrient loads leaching from the sites and the subsequent hydrogeochemical processes in groundwater. The piezometers (PVC) were installed after drilling the wells using a 2-inch Edelman hand auger. The final depth of the wells was reached after striking the stiff clayey layer, which was identified from initial investigations and previous studies (e.g. Kulabako et al., 2007) as the bottom confining layer of the shallow aquifer. This layer was difficult to penetrate by hand auger and indicated presence of consolidated clayey soils in the bottom layer. The shallow aquifer had a clayey-sand texture that was usually struck at depths ranging from 0.8 to 1.2 m below the ground level. Depending on the thickness of the aquifer, the piezometer screen length ranged from 0.5 to 1.5 m, but usually 0.5 m. The total depth of the wells varied from 1 to 3 m below ground level. The wells were capped at the bottom, and a nylon filter cloth wrapped around the screen used to filter out silts and clays. The wells were surveyed and levelled with respect to a permanent benchmark of known elevation. We observed that the aquifer in the study area was of alluvial type made up of clayey sand and with a very shallow water table (< 1 m).

4.3.2. Soil sampling, profiling and analysis

Soil samples were collected during the augering of wells. For each well, a number of soil samples were collected in black polyethylene bags for each soil layer or profile. The soil profiles were identified by examining the texture (by touching) and by visually inspecting the colour of the soil samples. The depths of these profiles and the water table were recorded. Samples were air dried, sieved using a 2 mm sieve and taken to the soil science laboratory of Makerere University (Kampala, Uganda) for analysis of pH, particle size distribution, available phosphorus and available Fe. The pH was determined from an aqueous soil suspension. Particle size distribution was determined with the Bouyoucos hydrometer method (Bouyoucos, 1962). Available phosphorus (adsorbed) was determined using the Bray 2 method (Bray and Kurtz, 1945), and available Fe was determined using the EDTA extraction method and by aspirating the filtrate into an atomic absorption spectrometer (Perkin - Elmer 2380) (Borggaard, 1979).

4.3.3. Groundwater sampling and analysis

Groundwater was sampled bi-weekly from March 2010 to August 2010. A foot pump fitted with a 1/2-inch diameter HDPE pipe was used to extract the samples. Before sampling, at least 3 well volumes were pumped out of the piezometer to ensure that the sample was representative of the aquifer water quality. Some wells were pumped a day before sampling because of the low recovery rates (typically 1 - 4 h to return to original water levels). Groundwater samples were collected in clean 500 ml polyethylene bottles. Immediately after taking the samples, pH, Electrical Conductivity (EC), temperature, Dissolved Oxygen (DO) and alkalinity (HCO_3^-) were determined in situ with calibrated WTW 3310 portable meters and a field-kit (for alkalinity). The samples for laboratory analyses were kept in a cool box at 4 °C and transported immediately to the Public Health and Environmental Engineering laboratory of Makerere University (2 km away from the study area). Here, the samples were analysed for nitrate (NO_3^-), ammonium (NH_4^+), orthophosphate (PO_4^{3-}) and total phosphorus (TP). NO_3^- was determined using the cadmium reduction method, NH_4^+ using the Nessler method, PO_4^{3-} using the Ascorbic acid method and TP using the Ascorbic acid method after digestion with persulphate (APHA/AWWA/WEF, 2005). Final readings were carried out on a HACH DR/4000 U spectrophotometer (USA). A few samples (a total of 40) were also analysed for Total Kjeldahl Nitrogen (TKN) using the Micro Kjeldahl method

(APHA/AWWA/WEF, 2005), whereby organic N was determined as the difference between TKN and ammonia N. For quality assurance, the procedures used were checked by measuring concentrations of known standards. Samples for cations (Ca, Mg, Na, K, Fe and Mn) and anions (Cl^- and SO_4^{2-}) were each filtered in the field in 50 ml scintillation vials, using a 0.45 mm cellulose acetate filter (Whatman), and kept cool at 4 °C in a cool box. Samples for cations were acidified by adding 2 drops of undiluted nitric acid to the scintillation vials to prevent precipitation and changes in cation concentrations. The scintillation vials were shipped to UNESCO-IHE, Delft, The Netherlands, for hydrochemical analyses using an Inductively Coupled Plasma spectrophotometer (ICP - Perkin Elmer Optima 3000) for cations and by Ion Chromatography (IC - Dionex ICS-1000) for anions. Dissolved Organic Content (DOC) and Silica (Si) were also determined from the acidified samples using a Total Organic Carbon analyser and the ICP, respectively.

4.3.4. Hydraulic conductivity and groundwater discharge

In order to compute discharge from groundwater levels, the saturated hydraulic conductivity (K) of soils in Bwaise was determined. Slug tests are widely used for the in situ determination of K because they can be performed quickly and at a low cost and, therefore, a good alternative to most other tests (Butler et al., 2003b; Hinsby et al., 1992). In this study, slug tests, which were based on Bouwer (1989) for fully penetrating wells, were performed at two locations at the waste dump site and at three locations at the pit latrine site to estimate K (m/d) of the aquifer. A screened PVC pipe was inserted in a freshly drilled 3 inch borehole and the water level in the pipe was allowed to stabilize, and then measured. A slug was then removed from the well using a bailer and a mini-diver (Eijkelkamp, the Netherlands) was immediately installed in the well to record the subsequent rise in water level. All slug tests were carried out in duplicate. K was determined using Eqns. (4.1) and (4.2) (Bouwer, 1989). To quicken the procedure for determining the constant C, a polynomial equation (Eqn. (4.3)), developed by Yang and Yeh (2004) was used.

$$K = \frac{r_c^2 \ln(R_e/r_w)}{2L_e} \frac{1}{t} \ln \frac{y_0}{y_c} \qquad (4.1)$$

$$\ln \frac{R_e}{r_w} = \left[\frac{1.1}{\ln(L_w/r_w)} + \frac{C}{L_e/r_w} \right]^{-1} \qquad (4.2)$$

$$C(x) = 1.605 + 9.496x - 12.317x^2 + 6.528x^3 - 0.986x^4 \qquad (4.3)$$

where L_e is the screen length (m), r_w is the radial distance of the undisturbed portion of the aquifer from the centerline which is also the radius of the auger (m), r_c is the radius of the casing pipe (m), L_w is the static water level from the bottom of the well (m), y is the vertical distance between the static water level and the water level inside the well (m) and t is the time (s). $y = y_0$ at time zero and $y = y_t$ at time t. R_e is the effective radius (m) of the well over which y_t is dissipated. The term x in Eqn. 3 is equal to Log (L_e/r_w).

Groundwater flow rates (m/d) were calculated using the Darcy's flow equation where the hydraulic gradient was determined from the water table contours at a given site. The groundwater discharge per unit width of flow was computed by multiplying the groundwater flow rate with the thickness of the aquifer at a given site (determined during augering of wells).

4.3.5. Data analysis and estimation of nutrient loads

Hydrochemical data were summarized using Stiff diagrams, and, in combination with chloride concentrations and nitrogen to chloride ratios (N:Cl), were used to assess the impacts of the studied sites on groundwater. The impacts were further assessed by comparing the average hydrochemical concentrations and nutrient loads upgradient and downgradient of the two study sites. A two sample students' t-test performed using SPSS statistical program was used to determine whether there was a significant ($p < 0.05$) difference between the upgradient and downgradient concentrations. Nutrient (NO_3^-, NH_4^+ and PO_4^{3-}) loads were estimated using a technique described by Jarvie et al. (2005) and Withers et al. (2011), which utilizes the product of the flow-weighted mean concentration and the average discharge/flow over the study period. This technique has a relatively small bias and variance compared with other techniques (Johnes, 2007; Webb et al., 1997). The loads were estimated from Eqn. (4.4). Using average groundwater flow rates, Eqn (4.4) was simplified to Eqn (4.5), which was used for the actual load estimations.

$$Load = k_r \left(\sum_{i=1}^{n} (C_i Q_i) / \sum_{i=1}^{n} Q_i \right) Q_{avg} \tag{4.4}$$

$$Load = k_r . C_{avg} . Q_{avg} \tag{4.5}$$

where C_i is concentration at the time of sampling (mg/L), Q_i is the groundwater discharge at the time of sampling (m³/s), Q_{avg} is the average groundwater discharge over the period of study (m³/s), C_{avg} is the average concentration over the period of study (mg/L) and k_r (-) is the conversion factor to take into account units of the load.

The processes potentially affecting dissolved nutrients (NO_3^-, NH_4^+ and PO_4^{3-}) transported into groundwater were identified based on the redox potential of the aquifer and the saturation indices (SI's) of compounds known to control phosphorus species. The redox potential (pe) of the aquifer was determined using the NO_3^-/NH_4^+ redox couple (Eqn. 4.6; Appelo and Postma, 2007):

$$pe = 14.9 - 1.25\, pH + 0.125 \log\left([NO_3^-]/[NH_4^+] \right) \tag{4.6}$$

The redox environment was also characterized from the presence or absence of major redox-sensitive parameters (including Fe^{2+}, Mn^{2+}, NH_4^+, NO_3^- and DO) based on the criteria developed by Stuyfzand (1993).

The PHREEQC code (Parkhurst and Appelo, 1999) was used to perform mineral speciation and saturation index (SI) calculations, expressed as $SI = -\log (IAP/K_{sp})$ where IAP and K_{sp} are the ion activity (-) and solubility product (-) constant of the mineral species respectively.

4.4. Results

4.4.1. Soils

The soils at the studied sites (section 4.3.1) contained alluvial material with three distinct layers: 1) a top layer containing reddish loam clay soils mixed with fill material; 2) a middle layer containing essentially grey sands with small amounts of clay; and 3) a bottom layer containing stiff clay soils with fine sand, grey-whitish in color (Fig. 4.4).

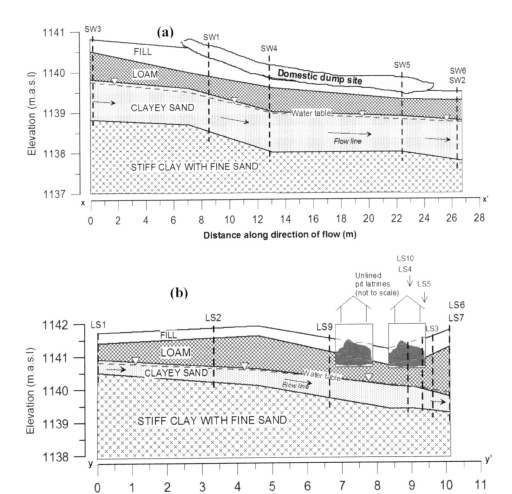

Figure 4.4: A simplified hydrogeological section across (a) the domestic solid waste dump site and (b) the pit latrine site.

Table 4.1 presents the average properties of these soil profiles. The middle layer had the highest sand content and formed the aquifer whereas the upper and lower layers acted as (semi) confining layers. The base of the sandy aquifer contained sand with brownish and rusty deposits, which were probably pockets of oxidised iron. The oxidation of Fe indicated that the aquifer was exposed to oxidized conditions and that it dried up frequently. From the data, we infer that the depositional environment of these sediments was likely low energy and associated with lakes or wetlands. Available P and Fe were generally low (< 1% by weight; Table 4.1) in the top, middle, and lower soil layers. The pH of the soils was slightly acidic to neutral (pH = 5.3 - 6.7; Table 4.1).

Table 4.1: Average physical and chemical properties of the different soils layers in the study sites.

Layer	Depth (m)	pH (-)	Sand (%)	Clay (%)	Silt (%)	Avail. P (mg/kg)	Fe (mg/kg)
Top (n=14)	0-0.9	6.7 ± 1.0	59 ± 5	28 ± 5	14 ± 2	18 ± 15	1157 ± 605
Middle (n=18)	0.9-1.8	6.0 ± 1.0	73 ± 7	17 ± 8	10 ± 2	10 ± 8	275 ± 216
Bottom (n=16)	> 1.8	5.3 ± 1.0	50 ± 7	42 ± 7	8 ± 3	6 ± 2	240 ± 123

4.4.2. Groundwater levels and flow direction

The water table was consistently encountered in the middle sandy layer at depths ranging from 0.5 to 1m below the surface. This layer formed the shallow aquifer of the study sites. The average thickness of the aquifer was 0.5 m at the pit latrine site, and 0.8m at the dump site (Fig. 4.4). The water table decreased gradually from April to August at both sites (Fig. 4.5), reflecting groundwater discharge and potentially evaporation losses in excess of rainfall. Fig. 4.1 presents the average water table contours at the two studied sites. The groundwater flow direction was consistent throughout the study period. The flow direction at both sites was approximately NE to SW, and thus, it was possible to distinguish between the upgradient and the downgradient groups of wells (Fig. 4.1).

4.4.3. Groundwater discharge

The saturated hydraulic conductivity (K) of the aquifer ranged from 0.05 to 2.9 m/d at the waste dump site and from 0.14 to 1.7 m/d at the pit latrine site (Table 4.3). The hydraulic gradient at the domestic waste dump site and the pit latrine site was approximately 0.04 and 0.05, respectively (Fig. 4.1), while the aquifer thickness was 0.8 m and 0.5 m, respectively (see section 4.4.2 and Fig. 4.4). Applying the Darcy equation, the average groundwater discharge per unit width at the domestic waste dump site ranged from 0.002 to 0.093 m^2/d whereas at the pit latrine site, it ranged from 0.004 to 0.043 m^2/d.

4.4.4. Hydrochemistry and nutrient processes

Groundwater beneath the two sites was contaminated with elevated concentrations of aqueous Na^+, Cl^-, NO_3^-, NH_4^+, SO_4^{2-}, DOC and HCO_3^- (see Table 4.2 and stiff diagrams in Fig. 4.1). Well SW2 was omitted from statistical analyses because it was suspected to be impacted by an adjacent pig stall (Fig.4.1). Wells downgradient of the pit latrine site, however, had higher concentrations of contaminants compared to the upgradient wells (Table 4.2 and Fig. 4.1). For example, the average EC downgradient of the pit latrine site was 5061 ± 1073 µS/cm compared to 1306 ± 306 µS/cm recorded at upgradient wells. Likewise, HCO_3^- increased from 322 ± 117 mg/l upgradient to as high as 1114 ± 356 mg/L downgradient of the pit latrines. The difference between the upgradient and downgradient concentrations of most hydrochemical parameters at the pit latrine site was statistically significant ($p < 0.05$) with the exception of Mg ($p = 0.4$: likely to be a natural source). Surprisingly, the considerably higher average PO_4^{3-} concentration downgradient of the pit latrines compared with the upgradient (Table 4.2) was not statistically significant ($p = 0.078$). This result suggested that ortho-phosphate was likely regulated by several processes in the aquifer. From the above, we conclude that wastewater, originating from the pit latrines, infiltrated the aquifer to alter water quality.

At the waste dump site, there was no significant difference ($p > 0.05$) between the upgradient and downgradient concentrations of most parameters, except for Ca ($p = 0.014$) and DOC ($p = 0.016$). There was a decrease in Ca^{2+} downgradient of the waste dump (probably due to ion exchange or precipitation) whereas DOC increased slightly suggesting moderate contamination of groundwater from the waste dump. There was a negligible difference between mean Cl^- concentrations (a conservative tracer) upgradient and downgradient of the dump site (Table 4.2) suggesting that the impact from the waste dump to groundwater was not significant. By contrast, the mean Cl^- concentration upgradient and downgradient of the pit latrine site was significantly different ($p = 0.001$).

Using the NO_3^-/ NH_4^+ redox couple, the redox potential (and pe) was found to range from Eh = 330 - 425mV (*pe* = 5.6-7.2). The redox environment was therefore sub-oxic ($0 \leq pe \leq 10$; Appelo and Postma, 2007) and we assert that nitrification, or the conversion of NH_4^+ to NO_3^-, was most likely taking place in the shallow aquifer. This would explain the general absence of Fe and the moderate levels of oxygen (about 2 mg/L) (Table 4.2). However, the co-existence of NH_4^+ and NO_3^- under these conditions (see Table 4.2) implies that hydrochemical heterogeneity with oxic and anoxic groundwater was present in close proximity to each other, especially at the pit latrine site. Despite the observed redox disequilibrium, the results generally showed that the aquifer conditions were not strongly oxidising as reflected by the limited presence of mobile Mn (Table 4.2).

Ortho-phosphate (PO_4^{3-}) was the dominant form of P in groundwater (60-80% of total P). Low concentrations of PO_4^{3-} (0.05 - 0.3 mg/L) were generally observed at both sites except downgradient of the pit latrines where average PO_4^{3-} concentrations were relatively high (2.4 mg PO_4^{3-}/L) (Table 4.2).

The saturation indices (SI) for vivianite ($Fe_3(PO_4)_2 \cdot 8(H_2O)$) and strenghite ($FePO_4$), both known to control P in groundwater, were negative at both sites (see Table 4.2) implying that the aquifer was undersaturated with respect to Fe minerals. The calculated indices suggested that iron phosphates were not likely potential sinks for PO_4^{3-}. It was also unlikely that the sorption of PO_4^{3-} to Fe oxyhydroxides on soil grain surfaces was significant in the removal of P from groundwater since Fe was almost absent in the soils (< 1% by weight; Table 4.1). Kulabako et al. (2008) carried out P isotherm experiments in the same study area and also found out that the Fe content of the soils (which was less than 1%) did not influence P sorption. Groundwater downgradient of the pit latrines was supersaturated with respect to hydroxyapatite ($Ca_5(PO_4)_3OH$) (SI > 0, Table 4.2). Likewise, both sites were supersaturated with respect to manganese hydrogenphosphate ($MnHPO_4$). This supersaturation suggested that $Ca_5(PO_4)_3OH$ and $MnHPO_4$ were potential sinks for PO_4^{3-} through precipitation. As noted by Reddy et al. (1999), Ca is one of the dominant species that governs PO_4^{3-} in soils that are neutral to alkaline, as in this study (section 4.4.1).

Table 4.2:   Average chemistry of shallow groundwater upgradient and downgradient of the pit latrine and the domestic solid waste dump sites. In brackets are the standard deviations. For TKN, n =10. Values are given in mg/L except for EC (Electrical conductivity, µS/cm), pH (-), T (Temperature, °C) and SI (-).

Parameter	Latrine site		Domestic waste dump site	
	Upgradient (n=4x5)	Downgradient (n=6x5)	Upgradient (n=4x5)	Downgradient (n=2x5)
EC	1306 (±306)	5061 (±1073)	859 (±141)	809 (±242)
pH	6.6 (±0.3)	7.5 (±0.2)	6.3 (±0.3)	6.5 (±0.2)
T	25 (±2)	25 (±2)	26 (±1.4)	25 (±1)
DO	2.3 (±0.9)	2.5 (±0.9)	2.8 (±1.7)	2.1 (±0.7)
Na^+	48 (±19)	172 (±114)	37 (±20)	30 (±13)
K^+	42 (±28)	307 (±173)	22 (±17)	51 (±48)
Mg^{2+}	16 (±21)	21 (±13)	6 (±4)	6 (±3)
Ca^{2+}	22 (±10)	61 (±32)	25 (±16)	12 (±5)
Fe^{2+}	0.02 (±0.02)	0.1 (±0.06)	0.04 (±0.06)	2 (±2)
Mn^{2+}	0.9 (±0.7)	0.4 (±0.24)	0.8 (±0.5)	0.7 (±0.4)
SO_4^{2-}	20 (±6)	127 (±86)	15.3 (±8.3)	9 (±8)
Cl^-	110 (±45)	380 (±206)	67 (±20.7)	69 (±35)
HCO_3^-	322 (±117)	1114 (±356)	171 (±71)	206 (±115)
NO_3^-	85 (±38)	228 (±237)	69 (±36)	24 (±17)
NH_4^+	4.9 (±5.1)	57 (±42)	2.9 (±1.7)	4 (±4)
PO_4^{3-}	0.05 (±0.09)	2.4 (±3)	0.3 (±0.6)	0.3 (±0.6)
TP	0.27 (±0.23)	3.2 (±11.3)	0.5 (±0.7)	0.47 (±0.3)
SiO_2	2.1 (±3.1)	4 (3.8)	1.1 (±1.9)	0.7 (±0.9)
DOC	5 (±3)	21 (±9)	2.8 (±1.4)	5 (±2.8)
TKN	4.9 (±0.9)	130.4 (±126)	8.8 (±6.8)	4.8 (±1.6)
*N:Cl (mg/l) ratio	0.2	0.25	0.27	0.1
** *Saturation indices (SI)*				
SI Hydroxyapatite	-6.4	1.2	-8	-6.7
SI Vivianite	-8.2	-5.1	-7.4	-2.6
SI $MnHPO_4$	1.2	2.2	2.0	-

* N in the N:Cl ratio is a combination of NH_4^+ (as N) and NO_3^- (as N) because both were dominant in the shallow groundwater at the two sites. Well SW2 was omitted from statistical analyses because it was impacted by the pig stall (Fig.1).

4.4.5. Temporal changes in nutrient concentrations at the pit latrine site

There was a consistent temporal trend in nutrients, especially NO_3^-, NH_4^+, and Cl^- which appeared to be associated with changes in groundwater levels and the leaching of wastewater from the pit latrine (Fig. 4.5).

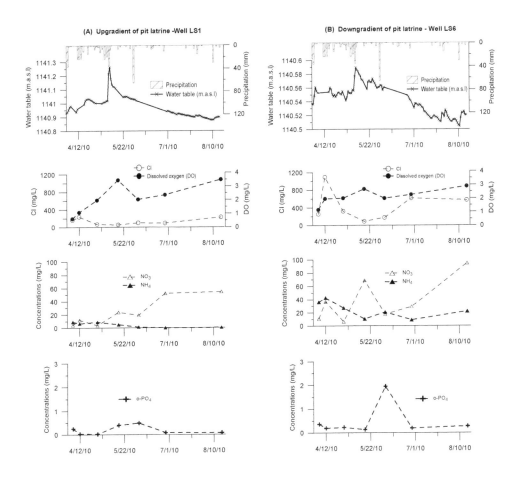

Figure 4.5:   Temporal variations in nutrient chemistry; (A) Upgradient of the pit latrine site, well LS1 and (B) downgradient of the pit latrine site, well LS6.

Between April and May (wet season), groundwater levels were high and an increase in NH_4^+ leached into groundwater was observed especially at the downgradient side of the pit latrine. From June through August (dry season), the water levels gradually dropped and NH_4^+ concentrations decreased whereas NO_3^- concentrations increased, suggesting a more oxidizing aquifer in the dry season and that nitrification was likely taking place. In the wet season, when water levels were highest, we also observed a decrease in NH_4^+ concentrations. Since Cl^- concentrations decreased concurrently, we attributed this decrease to dilution, since we regarded chloride as a conservative tracer whose concentrations can only be affected by dilution and practically insensible to other mechanisms (Fenton et al., 2009; Lyngkilde and

Christensen, 1992). Fig. 4.5 further shows that downgradient of the pit latrine, there was an increase in NH_4^+ concentrations later in the dry season, which was consistent with the continuous leaching of wastewater from the pit latrine. PO_4^{3-} concentrations were generally low throughout the wet and dry seasons (≤ 5 µmol/L) and only increased in early June in association with the intense short rains in that period (Fig. 4.5). Similar observations were made by Kulabako et al. (2008) in the same area whereby P concentrations in the shallow groundwater increased with rain events and occurrence of floods. This increase in groundwater P concentrations in the rainy season was attributed to leaching of P from the pit latrine caused by infiltrating rain water.

4.4.6. Nutrient loads

The calculated nutrient loads (NO_3^-, NH_4^+ and PO_4^{3-}) are summarized in Table 4.3. The PO_4^{3-} load transported in groundwater in the shallow aquifer generally ranged from 0 to 0.03 g/d. Downgradient of the pit latrine, however, the PO_4^{3-} load was higher and ranged from 0.01 to 0.1 g/d. At the waste dump site, apparently no significant ($p = 0.96$) P load was leached into groundwater. The N load transported in groundwater, on the other hand, was much higher than the P load (Table 4.2). Up to 0.9 - 9.8 g/d of NO_3^- was discharged daily from the pit latrine whereas the NH_4^+ load was in the range of 0.2 - 2.5 g/d. Downgradient of the waste dump, the NO_3^- and NH_4^+ loads were in the range of 0.05 – 2.23 and 0.008 – 0.37 g/d respectively. We concluded that owing to the presence of the pit latrine, there was a 168% increase in NO_3^- load, a 1050% increase in NH_4^+ load and 4900% increase in PO_4^{3-} load. In contrast, there was a 65% decrease in NO_3^- load, 35% increase in NH_4^+ load and 0% change in PO_4^{3-} load at the domestic waste dump site.

Table 4.3: Estimates of aquifer hydraulic conductivities (K; n=4) and nutrient loads (per unit width) in the plume from the pit latrine and the solid waste sites.

Study site	Range of K (m/d)	Location	Well groups	NO_3^- load (g/d)	NH_4^+ load (g/d)	PO_4^{3-} load (g/d)
Pit latrine	0.14 - 1.7	Upgradient	LS1, LS2, LS9, LS10	0.34 - 3.66	0.02 - 0.21	0 - 0.002
		Downgradient	LS3, LS4, LS5, LS6, LS7, LS8	0.91 - 9.80	0.23 - 2.45	0.01 - 0.1
Waste dump	0.05 - 2.9	Upgradient	SW1, SW3, SW7, SW4	0.14 - 6.42	0.006 - 0.27	0 - 0.03
		Downgradient	SW5, SW6, SW2	0.05 - 2.23	0.008 - 0.37	0 - 0.03

Concentrations used to calculate the nutrient loads were derived from Table 4.2. Well SW2 was omitted from the statistical analyses because it was believed to be impacted by the pig stall (see Fig.4.1)

4.5. Discussion

Although the results showed a high level of contamination at the studied sites, it was evident that substantial amounts of wastewater originated from the pit latrine and infiltrated the aquifer. As a result, nutrient concentrations downgradient of the pit latrines were high: 228 mg/L NO_3^-, 57 mg/L NH_4^+ and 2.4 mg/L PO_4^{3-}. Stiff diagrams in Fig. 4.1 further showed that changes in groundwater quality occurred at the pit latrine site. Pollution from the waste dump was substantially less than that from pit latrines. This difference may be attributed to attenuation of contaminants by physical, biological and chemical processes which could have occurred in the top unsaturated soil that separated the waste dump and the aquifer as noted by several researchers (e.g. Butler et al., 2003a; Christensen et al., 1994; Lyngkilde and Christensen, 1992). This explanation is in agreement with Biryabarema's (2001) findings which showed that there was only a slight increase in concentration of metals and ions downgradient of most solid waste dumps around Kampala city, which he partly attributed to presence of underlying lateritic soils that act as natural barriers to leachate migration. We also think that the nature of the material dumped at the site and the presence of polyethylene bags and plastic bottles (see section 4.3.1) could have acted as a pollutant seal beneath the waste dump even though there is currently no scientific evidence to support this explanation.

N to Cl ratios have been used to indicate anthropogenic pollution in groundwater (e.g. ARGOSS, 2002; Cronin et al., 2007; Flynn et al., 2012). The N:Cl (mg/L) ratios of the two studied sites ranged from 0.1 - 0.27 (Table 4.2), which were slightly lower but in close agreement with ratios calculated by Cronin et al. (2007) for polluted groundwater in the unsewered cities of Timbukutu in Mali (N:Cl = 0.3 to 1.9) and Lichinga in Mozambique (N:Cl = 0.5 to 3). The fact that the N:Cl ratio increased downgradient of the studied pit latrine whereas it decreased at the waste dump (see Table 4.2) supported our deduction that the pit latrine discharged its contents into groundwater, whereas there was comparatively limited discharge from the waste dump.

Generally, nitrogen (N) in wastewater is predominantly in the form of NH_4^+ or organic N. In this study, the dominant form of nitrogen in the shallow groundwater at the studied sites was inorganic (see Table 4.2) which implied that much of the organic N leached into the shallow aquifer had been converted to NH_4^+. Given the relatively high redox potential (pe = 5.6 - 7.2) of the studied sites, nitrification appeared to be the dominant process affecting NH_4^+ leached into groundwater. However, it was likely that full conversion of NH_4^+ to NO_3^- had not yet taken place since high concentrations of NH_4^+ (57 mg/L at the pit latrine and 5 mg/L at the domestic waste dump) were also measured. The most likely reasons for this could be:

1. Conditions were not strongly oxidising, because the redox environment in the aquifer was sub-oxic as explained in section 4.4.4,

2. Presence of redox disequilibrium beneath the contaminated pit latrine site due to the short travel distances, and

3. The continuous input of NH_4^+ mass from pit latrine contents via the ammonification process was higher than the rate of nitrification.

In relation to the latter point, it was indeed evident from Fig. 4.5 that NH_4^+ was continuously leached into groundwater. In the dry season, the aquifer was more oxidising and some NH_4^+ was converted to NO_3^-, whereas dilution likely took place in the wettest periods. Ortho-phosphate concentrations were generally low throughout the wet and dry seasons and were likely regulated by a number of processes in the aquifer, as described below.

Given that some mobile Mn^{2+} was measured, then the conditions in the shallow aquifer were not strongly oxidising, despite the relatively high redox potential. This observation also suggested that the redox conditions in the aquifer were Mn-reducing. Thus, Mn^{2+} could have had a potential to regulate P. Indeed the SI of $MnHPO_4$ was positive at both sites, whereas that of Fe-minerals was negative (Table 4.2). As described in section 4.4.4, the SI of $Ca_5(PO_4)_3OH$ beneath the pit latrine was also high and, therefore, the mineral likely regulated PO_4^{3-} transported in groundwater. Given that Fe^{2+} was absent in groundwater as well as in the soil (see Table 4.1 and Table 4.2), we concluded that iron did not play a major role in PO_4^{3-} retention in the soils. Most probably, P leached into groundwater was regulated by precipitating with Ca^{2+} and Mn^{2+} ions (e.g. Robertson et al., 1998; Zanini et al., 1998).

Results from this study further showed that the plume from the pit latrine discharged 0.91 - 9.8 g/d NO_3^-, 0.23 - 2.45 g/d NH_4^+ and 0.01 - 0.1 g/d PO_4^{3-} per unit width of flow (Table 4.3). Given a plume width of approximately 5 m with a length of about 1m at the point of measurement (Fig. 4.1), then the actual dissolved nutrient masses leached into groundwater were 1.03 - 11.14 g/d as NO_3-N, 0.9 - 9.56 g/d as NH_4-N and 0.02 - 0.16 g/d as PO_4-P. The total inorganic nitrogen (NO_3-N + NH_4-N) load leached into groundwater therefore ranged from 1.93 to 20.7 g/d. Each person in Uganda generates about 2.5 kg of N and 0.4 kg of P per year (Jönsson et al., 2004) which translates into 6.9g of N and 1.1g of P per day. With these estimates and the observed number of users of the toilets at the study site (15 users per day - personal observation), the actual nutrient input in the pit latrine was about 104g of N and 17g of P per day. A comparison of the nutrient mass input into the pit latrine and the groundwater system, suggested that approximately 2 - 20% of N input and 0 - 1% of P input, was leached to groundwater from the pit latrines. Hence, a substantial amount of nutrients was either retained in the pit latrines or in the soil between the latrine and the piezometers. P leached into groundwater was likely retained in soils by precipitating as $MnHPO_4$ and $Ca_5(PO_4)_3OH$, as discussed above. On the other hand, N, leached into groundwater predominantly as ammonium (NH_4^+), was converted to NO_3^- due to the relatively high redox potential of the aquifer (section 4.4.4). The NO_3^- ion is a fairly stable end product for N leached into groundwater under these oxic conditions. Another likely pathway through which N could have been lost from the pit latrine was by denitrification, where N in the pit is converted to N_2 (nitrogen gas), which is then lost via the vent pipe of the pit latrine. Denitrification is reported by several researchers as a major process in pit latrines (e.g. Jacks et al., 1999; Montangero and Belevi, 2007). The loss of N as volatile ammonia from groundwater was, however, likely negligible because of the neutral pH (pH = 6.6 - 7.5) and high temperature (T > 20°C), which favour low ammonia volatilisation (e.g. Jacks et al., 1999; Schilling, 2002).

Our results demonstrated that the unlined raised pit latrines, which were directly connected to the shallow groundwater system, had an influence on groundwater quality since concentrations of almost all ions increased significantly. From this, we conclude that the pit latrine site we studied polluted groundwater. However, at the same time, the removal of PO_4^{3-} was almost complete due to retention processes either in the two pit latrines or in the shallow subsurface in the immediate vicinity of the two latrines, whereas removal of N species was more than 80%. Although not within the scope of this study, the removal of N and P can also be a result of a number of biological processes such as the sulfur cycle-associated Enhanced Biological Phosphorus Removal (EBPR) process (Brdjanovic et al., 1998; Wu et al., 2014). If P is the limiting factor for primary production and eutrophication in local aquatic systems (e.g. Verhoeven et al., 1996), one possible implication of our findings is that many pit latrines may not be major contributors to local eutrophication compared to other sources of pollution (e.g. open defecation or grey water discharge in open channels) due to the almost complete retention of PO_4^{3-}. However, there are several points of caution for this interpretation. First of all, we investigated only two pit latrines and we do not know if and how our results can be upscaled. Second, there are many pit latrines in the Bwaise area. Although the percentage of nutrient N and P leached from the pit latrine into the aquifer system is small, the cumulative loading from all pit latrines could be substantial as was reported by Kulabako et al. (2004). Third, we have not determined the P buffering capacity of the shallow aquifer. The aquifer can become saturated with P due to the continuous leaching of wastewater from the latrines into groundwater, and the fact that there are so many pit latrines. This situation may eventually lead to increased mobile P concentrations that are potentially capable of causing eutrophication of surface water located downstream in the catchment. Fourth, the removal of nutrients is a function of geochemical conditions which may be different in other locations where pit latrines are used.

Given that pit latrines are the dominant mode of excreta disposal for Kampala City and other areas of sub-Saharan Africa, how may these systems be improved to reduce groundwater contamination and downstream eutrophication risks? As the studied waste dump was found not to significantly pollute groundwater, this finding can be used to suggest possible improvements to pit latrines, especially in areas with shallow water tables. First, given that most nutrients at the waste dump site were likely retained in the first top unsaturated soil, pit latrines can be elevated with the vault(s) above the ground surface to minimize the contact between the pit and the aquifer and to increase the thickness of the unsaturated zone between the bottom of the pit and the aquifer. Alternatively, an artificial barrier of fine loamy clayey soils (or lateritic soils) can be placed around the pit of newly constructed pit latrines to inhibit pollutant transport while allowing liquids to percolate. Similar measures have been proposed in the past. Franceys et al. (1992) for example suggested that groundwater pollution can be significantly reduced if there is at least a 2m layer of relatively fine soil between the pit and the water table whereas (Coetzee et al., 2011) suggested using a biogeochemical filter (packed with stones) placed beneath a pit to remove the bulk of nitrogen leached. Second, since the leachate production at the waste dump was likely low owing to the relatively low moisture content, this concept can be applied to pit latrines by minimizing the water content in the pit contents through the use of urine-diverting dry toilets (UDDT's). Urine can be allowed to drain in a soakaway filled with fine clayey loamy soil. UDDT's have been promoted in the past in Uganda but they still face challenges with ease of use and cultural acceptance. Other possible improvements include (1) use of additives (e.g. $Ca(OH)_2$) to increase pH of pit contents/leachate and thus increase loss of N through ammonia volatilisation and (2) use of bio-gas toilets that allow for volatilization of ammonia due to high temperatures of pit contents.

4.6. Conclusions

In this chapter, we investigated the hydrochemical processes and nutrient mass flows in groundwater originating from two common on-site sanitation systems in an urban slum in Kampala, Uganda: a traditional pit latrine site (with 2 adjacent pit latrines) and an open domestic solid waste dump site. We found that a significant amount of wastewater (and nutrients for that matter) was leached from the pit latrines to groundwater (228 mg/L as NO_3^-, 57 mg/L as NH_4^+ and 2.4 mg/L as PO_4^{3-}), whereas the domestic waste dump site did not have a substantial impact on shallow groundwater. Temporal variations in nutrient chemistry showed that N was leached into groundwater throughout the wet and dry seasons whereas P was only leached during intense rains.

We estimated that approximately 2 - 20% of N input and less than 1% of P input were leached to groundwater from the pit latrines. The bulk of N leached into groundwater was in the form of ammonium (NH_4^+) whereas for P, it was PO_4^{3-} (ortho-phosphate). Ammonia volatilization was found to be negligible. Nitrification (the conversion of NH_4^+ to NO_3^-) was the major process affecting NH_4^+ leached to groundwater, because of the relatively high redox potential of the aquifer. However, complete nitrification did not occur as indicated by the continued presence of NH_4^+ and the sub-oxic redox conditions which were not strongly oxidizing. The aquifer instead exhibited Mn-reducing conditions. P leaching into groundwater as PO_4^{3-} was potentially regulated by precipitation as manganese hydrogenphosphate ($MnHPO_4$) and hydroxyapatite ($Ca_5(PO_4)_3OH$).

In terms of managing eutrophication, our results indicated that the continued use of pit latrines might be a relatively good option for disposal of faecal sludge given that a substantial amount of nutrients (especially P) were either retained in the pit latrine or the surrounding soils adjacent to the pit latrine. However, we realize that the results from this study need to be upscaled considering that the cumulative nutrient loading from all pit latrines in the entire Bwaise area could be substantially higher than the findings from this study.

Based on findings from this study, some improvements to the current pit latrine systems are suggested, especially in areas with shallow water table. These include advocating for elevated pit latrines, installing a biogeochemical filter of fine loamy clayey soils around the pit, adopting urine diverting dry toilets (UDDTs) and using pit additives to increase loss of N by volatilization.

Chapter 5

Understanding the fate of sanitation-related nutrients in a shallow sandy aquifer below an urban slum area

Abstract

We hypothesized that wastewater leaching from on-site sanitation systems to alluvial aquifers underlying informal settlements (or slums) may end up contributing to high nutrient loads to surface water upon groundwater exfiltration. Hence, we conducted a hydro-geochemical study in a shallow sandy aquifer in Bwaise III parish, an urban slum area in Kampala, Uganda, to assess the geochemical processes controlling the transport and fate of dissolved nutrients (NO_3^-, NH_4^+ and PO_4^{3-}) released from on-site sanitation systems to groundwater. Groundwater was collected from 26 observation wells. The samples were analyzed for major ions (Ca^{2+}, Mg^{2+}, Na^+, Mg^{2+}, Fe^{2+}, Mn^{2+}, Cl^- and SO_4^{2-}) and nutrients (PO_4^{3-}, NO_3^- and NH_4^+). Data was also collected on soil characteristics, aquifer conductivity and hydraulic heads. Geochemical modeling using PHREEQC was used to determine the level of PO_4^{3-} control by mineral solubility and sorption. Groundwater below the slum area was anoxic and had near neutral pH values, high values of EC (average of 1619 µS/cm) and high concentrations of Cl^- (3.2 mmol/L), HCO_3^- (11 mmol/L) and nutrients indicating the influence from wastewater leachates especially from pit latrines. Nutrients were predominantly present as NH_4^+ (1 − 3 mmol/L; average of 2.23 mmol/L). The concentrations of NO_3^- and PO_4^{3-} were, however, low: average of 0.2 mmol/L and 6 µmol/L respectively. We observed a contaminant plume along the direction of groundwater flow (NE-SW) characterized by decreasing values of EC and Cl^-, and distinct redox zones. The redox zones transited from NO_3-reducing in upper flow areas to Fe-reducing in the lower flow areas. Consequently, the concentrations of NO_3^- decreased down-gradient of the flow path due to denitrification. Ammonium leached directly into the alluvial aquifer was also partially removed because the measured concentrations were less than the potential input from pit latrines (3.2 mmol/L). We attributed this removal (about 30%) to anaerobic ammonium oxidation (Anammox) given that the cation exchange capacity of the aquifer was low (< 6 meq/100g) to effectively adsorb NH_4^+. Phosphate transport was, on the other hand, greatly retarded and our results showed that this was due to the adsorption of P to calcite and the co-precipitation of P with calcite and rhodochrosite. Our findings suggest that shallow alluvial sandy aquifers underlying urban slum areas are an important sink of excessive nutrients leaching from on-site sanitation systems.

This chapter is based on the manuscript: Nyenje, P.M., Havik, J.C.N., Foppen, J.W., Muwanga, A. and Kulabako, R. 2013, Understanding the fate of sanitation-related nutrients in a shallow sandy aquifer below an urban slum area: Journal of Contaminant Hydrology, http://dx.doi.org/10.1016/j.jconhyd.2014.06.011."In Press"

5.1. Introduction

Nutrients released from wastewater leaching from on-site sanitation systems to groundwater can be of several orders of magnitude higher than the minimum required to cause surface water eutrophication (Nyenje et al., 2013a; Parkhurst et al., 2003; Robertson et al., 1998; Weiskel and Howes, 1992; Zurawsky et al., 2004) The principal nutrients that cause eutrophication are nitrogen (N) in the form of nitrate (NO_3^-) and ammonium (NH_4^+), and phosphorus (P) in the form of ortho-phosphate (PO_4^{3-}) (Correll, 1998; Thornton et al., 1999). Hence, several studies have attempted to understand the fate of wastewater-derived nutrients in shallow groundwater by assessing the geochemical and hydro-chemical processes governing their transport (e.g. Corbett et al., 2000; Griffioen, 2006; Navarro and Carbonell, 2007; Ptacek, 1998; Zanini et al., 1998) or by modeling these processes using reactive transport models like PHREEQC (e.g. Parkhurst et al., 2003; Spiteri et al., 2007; van Breukelen and Griffioen, 2004). These studies have shown that nutrients leaching to groundwater can undergo a number of transformations, which may regulate their transport depending on existing hydro-chemical conditions. This may in turn limit or increase the transport of nutrients and their impact to downstream surface water systems. For example, the transport of PO_4^{3-} in groundwater is highly retarded owing to its strong adsorption affinity to metal oxides (esp. Fe and Mn), the co-precipitation of P with calcite and the precipitation of P with phosphate minerals such as hydroxyapatite (Golterman, 1995; Robertson et al., 1998; Shenker et al., 2005). However, PO_4^{3-} can also be easily remobilized in anoxic groundwater following reductive dissolution of Fe-/Mn-oxides (e.g. Zurawsky et al., 2004) or by inhibition of precipitation reactions due to the presence of other minerals (e.g. Cao et al., 2007) or the presence of highly alkaline conditions (Reddy et al., 1999). N species are mainly regulated by redox processes. The main driver of these processes in groundwater is the dissolved organic carbon (DOC) (Appelo and Postma, 2007). Under aerobic conditions, N originally present as NH_4^+ in wastewater leachates can be readily oxidized to NO_3^-, which is relatively conservative and may end up in lakes causing eutrophication (e.g. Pieterse et al., 2005). In highly anaerobic conditions, however, N present as NO_3^- can be removed from the aquifer by denitrification (e.g. Rivett et al., 2008).

So far, little information is available on the role of the above mentioned processes in determining the fate of nutrients in shallow alluvial aquifers underlying unsewered urban informal settlements or slums. Such environments are characterized by densely-populated informal settlements usually without access to proper on-site sanitation systems (Enabor, 1998; Katukiza et al., 2012; Kulabako et al., 2010). Human excreta is predominantly disposed of via poorly constructed pit latrines whereas most wastewater is disposed, untreated, over compounds or directly in excavated pits and drainage channels (e.g. Enabor, 1998; Kulabako et al., 2010). A significant number of people in these settlements also lack access to toilet facilities and some practice open defecation. In Uganda for example, 1 - 2% of the urban population are still reported to engage in open defecation (WHO/UNICEF, 2013). These activities have been shown to heavily contaminate shallow groundwater with high levels of faecal coliforms and nutrients (e.g. ARGOSS, 2002; Cronin et al., 2007; Montangero and Belevi, 2007; Nyenje et al., 2013a). Most studies in slum areas and other peri-urban areas have focused on understanding and managing immediate health risks associated with the contamination of drinking water supplies (e.g. springs and wells) by pathogenic organisms (ARGOSS, 2002; Howard et al., 2003; Kimani-Murage and Ngindu, 2007; Kulabako et al., 2007; Lutterodt et al., 2014; Mkandawire, 2008). Nitrate contamination is also a widespread problem dealt with in many papers investigating nitrate-transport processes, but it has been mainly linked to agricultural practices (e.g. Andrade and Stigter, 2011; Stigter et al., 2011).

However, research on the transport of sanitation-related nutrients in shallow groundwater in slum areas is lacking. This constrains efforts of safeguarding surface water ecosystems from risks associated with eutrophication when nutrient-rich groundwater exfiltrates to surface water systems. In this study, we attempted to investigate the geochemical processes governing the transport and fate of sanitation-related nutrients (NO_3^-, NH_4^+ and PO_4^{3-}) in a shallow sandy aquifer in Bwaise III parish, a slum area in Kampala, Uganda. Specifically our objectives were to: (i) characterize the hydrogeology and geochemistry of the alluvial aquifer (ii) quantify the hydrochemistry, nutrients (P and N) and redox environment in the contaminated shallow groundwater, and (iii) establish the geochemical and hydro-chemical processes controlling the fate of wastewater-derived contaminants and nutrients in the alluvial aquifer.

5.2. Study area (Bwaise III parish slum)

Bwaise III parish (54 ha) is a poorly sanitized urban informal settlement (or slum area) located in a low-lying swampy area, North West of Kampala city, Uganda (Fig. 5.1). It has one of the highest population densities in the city (about 300 persons/ha). There are 5 administrative zones in the parish (Fig. 5.1). This research focused on two zones: St. Francis and Katoogo.

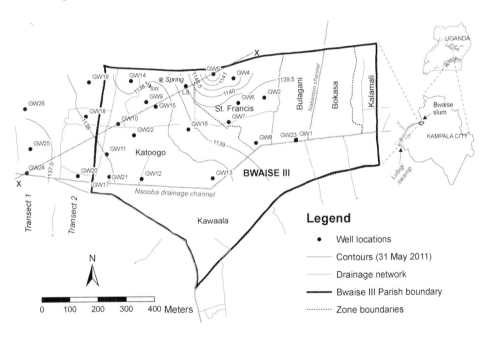

Figure 5.1:   Map showing the groundwater monitoring network in Bwaise III parish slum and transects; transect 1 (GW20, GW18 and GW19) and transect 2 (GW24, GW25 and GW26), which are located downstream the slum area near Lubigi swamp.

Details of the study area are already described in section 4.2 and 4.3.1. Here, we only provide additional information relevant to this particular study. As mentioned earlier (see chapter 4), the slum area is situated in a low-lying area with alluvial deposits, which are characterized by a shallow groundwater table ($<$ 1m) and clay soils overlain by fine sand and loamy top soils (Kulabako et al., 2007; Nyenje et al., 2013a). These deposits were derived from weathering and erosion from upper areas and hill summits. The geology in upper areas is characterized by a deeply weathered lateritic regolith (about 30m thick) overlying Precambrian basement rocks consisting of predominantly undifferentiated granite-gneiss rocks of the Buganda-Toro Cover Formation (Taylor and Howard, 1996, 1999b). Laterite is a highly weathered mineral rich in iron and aluminium oxides. X-ray diffraction analyses show that the weathered regolith is dominated by kaolinite ($Al_2Si_2O_5(OH)_4$) as the most abundant clay mineral and quartz with small amounts of iron oxide (Flynn et al., 2012). Iron oxides have been reported to be present mainly in the crystalline forms as goethite ($FeOOH$) (Biryabareema, 2001; Taylor and Howard, 1999b) and to a small extent as heamatite (Fe_2O_3) (Flynn et al., 2012). The shallow groundwater in the regolith aquifer is slightly oxic and rich in nitrate and usually discharges as springs or diffuses to shallow alluvial aquifers in the low-lying areas. The area experiences two rainfall seasons (March - May and September - December) with a mean annual rainfall of about 1450 mm/y.

Most residents in the study area have access to piped water supply, although some prefer to rely on springs to minimize costs (Kulabako et al., 2010). On-site sanitation is, however, very poor and as a result, shallow groundwater is heavily contaminated. Elevated concentrations of P (10 - 15 mg TP/L), N (15 - 30 mg TKN/L), Cl^- (40 – 80 mg/L), faecal coliforms (up to $28x10^8$ cfu/100mls) and high values of EC (700 - 5000 µS/cm) have therefore been reported in the shallow groundwater (Flynn et al., 2012; Kulabako et al., 2007; Nyenje et al., 2013b). This contamination is largely attributed to the widespread use of un-improved traditional pit latrines for excreta disposal, the practice of open defecation and flying toilets (use of polyethylene bags for excreta disposal), the poor solid waste management and the indiscriminate disposal of untreated grey water (Katukiza et al., 2012; Kulabako et al., 2007). Most contamination to groundwater is, however, believed to originate from pit latrines. This is because over 80% of the residents depend on pit latrines for excreta disposal. Only 1.3% use septic tanks (Katukiza et al., 2010b) and less 5% practice open defecation (CIDI, 2006). Hence, there is a very high density of pit latrines in the study area (average of 10 pit latrines per hectare) with the majority located in Katoogo zone (29% of the total) and St. Francis (24% of the total) (Kulabako et al., 2004). These pit latrines are usually unlined at the bottom implying that the contents leach directly into the shallow groundwater (see chapter 4). Additionally, due to the high water table ($<$ 1 m), most pit latrines are elevated (see Fig. 5.2). Consequently, the pits fill up very fast and the pit contents are usually emptied manually into nearby channels or adjacent pits. This practice is also attributed to the poor accessibility in the area and the high costs of pit emptying. Hence, the contamination of groundwater from pit latrines is usually very high and leachate concentrations as high as 2.4 mg/L for PO_4^{3-} and 57 mg/L for NH_4^+ have been reported (Nyenje et al., 2013a). The open disposal of grey water is another possible source of groundwater contamination in the study area. However, its contribution in relation to pit latrines is currently unknown.

Figure 5.2:   A typical elevated pit latrine in Bwaise III parish slum, Kampala Uganda. The pit latrines are usually elevated or raised above the ground due to the high water table in the area. This prevents the pits from filling up with water. Also see Fig. 4.3a

5.3.   Materials and methods

5.3.1. The monitoring network

The monitoring network consisted of 26 wells (Fig. 5.1). These included 19 wells installed in a randomly distributed pattern in Bwaise III parish slum in St. Francis and Katoogo zones, and 6 wells installed along two transects (each with 3 wells) in an uninhabited area downstream of the slum area (Fig. 5.1). The elevation of each well was determined with respect to a permanent benchmark (a reference point of known height above the mean sea level). The wells in the Bwaise III slum area were installed specifically to understand the chemical composition and the state of nutrients in the shallow groundwater. The well transects, on the other hand, gave an indication of the fate of ions and nutrients generated from the slum area.

The well holes were drilled using a 50 mm Edelman hand auger. The wells consisted of PVC pipes (50 mm internal diameter) connected to slotted screen pipes. The pipes were capped at the bottom to prevent sediments from entering the well. Each well had varying screen lengths (0.5 – 1 m) and pipe lengths (1 – 3 m) depending on the thicknesses of the aquifer and the overlying soil. The final depth of the wells was reached after striking the stiff clayey layer. This layer was difficult to penetrate by hand auger and indicated presence of consolidated clayey soils acting as the bottom confining layer of the shallow aquifer. A cloth made of a nylon silk material was wrapped around the screen before installing the wells to act as a filter material for silts and clays, which would otherwise clog the wells. This kind of installation also avoids changes in water quality entering the well due to possible cation exchange with the commonly used well materials such as the gravel pack (Appelo and Postma, 2007). A concrete base (equipped with a metallic cover lid) was constructed around the top of each well at the ground surface to avoid vandalism and contamination of the wells from surrounding human activities (refer to Fig. 4.2). After installation, the wells were developed by pumping them dry daily for a week until they reached full performance.

Groundwater levels were recorded once every week from February to May 2011. These levels were used to construct flow contours using the Kriging interpolation technique of the Surfer 8 software (Golden Software Inc., Golden, CO, USA). The contours were used to estimate the direction of flow and the head gradients.

5.3.2. Estimating hydraulic conductivity, K (m/d), and groundwater flow

Hydraulic conductivity measurements (K, m/day) were carried out using slug tests based on the Bouwer and Rice method (Bouwer, 1989). The application of this test including the step by step field procedure followed and the equations used to calculate K, have been previously described chapter 4 (section 4.3.4) These tests were carried out at 5 selected sites (GW16, GW1, GW4, GW10 and SW) (Fig. 5.1) evenly distributed across the study area. At each site, the tests were carried out in duplicate. Using the average K value and the hydraulic gradient estimated from the water table contours, we were able to estimate the groundwater flow velocity (m/day) using the Darcy equation.

5.3.3. Soil sampling and analysis

A total of 56 soil samples were collected during the augering of well holes. The soil samples were collected from each distinctive layer encountered in the borehole profile. Distinctions between soil layers were made by examining the texture (by touching) and by visually inspecting the color of the soil samples. The samples were kept in black polyethylene bags and transferred to Makerere University Public Health and Environmental Engineering Laboratory, Kampala Uganda for drying and initial analyses, and then to UNESCO-IHE Laboratory, The Netherlands for further analyses. At Makerere University, the samples were air dried for 2 weeks, sieved through a 2 mm sieve and analyzed for particle size distribution using the hydrometer method (Bouyoucos, 1962), pH measured on a 2.5:1 water-soil suspension, organic carbon (OC) using the Walkley–Black method (Walkley and Black, 1934) and cation exchange capacity (CEC, meq/100g) estimated by summing the soil extractable cations (Ca^{2+}, Mg^{2+}, K^+ and Na^+) extracted using the ammonium acetate method (Schollenberger and Simon, 1945). At UNESCO-IHE, geo-available metals (especially the P-adsorbing metals: Fe, Ca, Al and Mg) were extracted with 0.43 M HNO_3 (Novozamsky et al., 1993; Rauret, 1998) and analyzed using an Inductively Coupled Plasma spectrophotometer (ICP – Perkin Elmer Optima 3000). Geo-available metals here refer to the long-term mobilizable metals (Pettersen and Hertwich, 2008).

5.3.4. Groundwater sampling and analysis

Groundwater samples were collected from the monitoring wells using a portable hand pump for 2 – 3 times a month between February and May 2011. Before sampling, the wells were purged at least three (3) well volumes to ensure that the samples obtained were fresh and representative of site conditions. For some wells, where recovery times were very low (typically 4 – 6 hours), they were purged a day before. This approach has also been applied in other similar studies (e.g. Carlyle and Hill, 2001). We also collected spring water, which represented the inflowing shallow groundwater originating from the upper regolith aquifer (see Chapter 3).

Groundwater was collected in clean 500 ml polyethylene bottles. Immediately after sampling, the following parameters were determined on-site: EC, pH, DO, Alkalinity (HCO_3^-), NH_4^+, PO_4^{3-} and NO_3^-. EC and temperature were measured with an EC electrode (TetraCon 325,

WTW) connected to an EC meter (WTW 3310), pH with a pH electrode (SenTix 21, WTW) connected to pH meter (WTW 3310) and DO with a DO sensor (CellOx 325, WTW) connected to a DO meter (WTW 3310). All meters were calibrated before taking measurements. HCO_3^- was determined by titrating with 0.2 M sulphuric acid. A portable colorimeter (Hach DR890) was used to determine PO_4^{3-}, NH_4^+ and NO_3^- on-site using standard Hach protocols. NO_3^- was determined using the cadmium reduction method, NH_4^+ using Nessler's method and PO_4^{3-} using the ascorbic acid method. Samples for cations (K^+, Mg^{2+}, Ca^{2+}, Mg^{2+}, Fe^{2+} and Mn^{2+}) and anions (Cl^- and SO_4^{2-}) were filtered on-site through a 0.45 µm membrane filter into 25 ml scintillation vials and then kept cool at 4°C. Cation samples were further preserved by adding two drops of conc. nitric acid. These samples were shipped to UNESCO-IHE analytical laboratory, Delft, The Netherlands, for determination of cations using an Inductively Coupled Plasma spectrophotometer (ICP - Perkin Elmer Optima 3000) and anions by Ion Chromatography (IC - Dionex ICS-1000). Dissolved Organic Carbon (DOC) was also determined from the acidified samples using the Total Organic Carbon analyser (Shimadzu, TOC-V PCN).

Some groundwater samples were unstable and changed to brown color precipitates within 5 minutes after sampling. Hence, on-site analyses and filtrations were always done as fast as possible (< 5 minutes) to avoid possible changes in water quality (e.g. precipitation of Fe^{2+} and Ca^{2+} or conversion of NH_4^+ to NO_3^- etc.). Given the complexity of collecting and preserving water samples in slum environments, we considered an ion charge balance error of 20% to be sufficient for this study.

5.3.5. Hydrochemical data analysis

The processes likely governing the transport and transformation of nutrients were assessed by investigating the hydrochemistry and geochemistry of the shallow aquifer, the groundwater flow and the nutrient transport processes. The hydrochemistry of the shallow groundwater was characterized using the iso-concentration maps, redox sensitive species and the water quality classification based on the Stuyfzand classification method (Stuyfzand, 1989; Stuyfzand, 1993). The Stuyfzand method was preferred to the traditional piper diagram because it takes into account rare water types such as $Ca-NO_3$ and NH_4-HCO_3, which were typical of the studied shallow groundwater. The iso-concentration maps were used to analyze the changes in hydrochemistry along the groundwater flow path. Geochemical controls on nutrient transport were assessed by geochemical modelling and redox characterization.

5.3.6. Geochemical modelling

Mineral dissolution and precipitation

The PHREEQC code (version 2) (Parkhurst and Appelo, 1999) was used to carry out geochemical speciation and to calculate the saturation indices (SI's) of phosphate minerals in order to predict the mineral phases likely controlling the presence of PO_4^{3-} in groundwater (Appelo and Postma, 2007; Deutsch, 1997). Phosphate minerals considered were vivianite ($Fe_3(PO_4)_2.8H_2O$), hydroxyapatite ($Ca_5(PO_4)_3.OH$) and manganese hydrogen phosphate ($MnHPO_4$). Strengite was not considered because the aquifer was anoxic (see section 5.4). Carbonates of Fe, Ca and Mn are also known to significantly regulate PO_4^{3-} concentrations by co-precipitation or by regulating metal ions via the common-ion effect (Deutsch, 1997; Reddy et al., 1999; Spiteri et al., 2007). Hence, the SI's of siderite ($FeCO_3$), calcite ($CaCO_3$) and rhodochrosite ($MnCO_3$) were also considered. The SI's of Fe oxides and hydroxides (e.g. hematite) were not calculated because they have slow Kinetics (Deutsch, 1997) and were therefore unlikely to regulate PO_4^{3-} concentrations in the shallow groundwater.

Phosphate adsorption on the aquifer material

The PHREEQC code was also used to model the adsorption of ortho-phosphate (including the forms: PO_4^{3-}, HPO_4^{2-}, $H_2PO_4^-$) onto Fe-oxides using a surface-complexation process (Dzombak and Morel, 1990; Parkhurst and Appelo, 1999). We considered Fe-oxides to be the most capable of adsorbing P because Fe was one of the most dominant geo-available metal (as shown later on in results; Table 5.1) and because Fe oxides/hydroxides are considered as the principal sorbent of phosphorus in most aquifers and in the PHREEQC model code. The sorption process proceeds by the following reaction (Parkhurst et al., 2003):

$$SiteOH + PO_4^{3-} + 2H^+ = SiteHPO_4^- + H_2O \tag{5.1}$$

Where Site represents the phosphorus sorption site (e.g. Fe-oxide). The equilibrium constant (*log K_{sp}*) for the above reaction depends on the type of the sorbing surface. In this study, we used the default database in the PHREEQC model code, which contained thermodynamic data for the surface called hydrous ferric oxide (Hfo, or ferrihydrite), derived from Dzombak and Morel (1990). Two binding sites are defined in the database: a strong binding site (Hfo_s) and a weak binding site (Hfo_w). These sites were characterized by Dzombak and Morel (1990) as follows: 0.2 mol weak sites/mol Fe, 0.005 mol strong sites/mol Fe, a surface area of 600m²/g Fe and a molar weight of 89 g Hfo/mol. The type of Fe-oxide used in this particular study was goethite because it is the most abundant Fe-oxide in the study area (see section 5.2). Goethite has a surface area, which is about 10 times less than the default hydrous ferric oxide used in the PHREEQC model code (Appelo and Postma, 2007). We therefore assumed 10 times fewer sorption sites in the model, i.e. 0.02 mol weak sites/mol Fe, 0.0005 mol strong sites/mol Fe and a specific area of 60 m²/g. The total amount of iron oxide present was specified as the amount of Fe extracted using the 0.43 M nitric acid extraction method. In the section 5.4, the model results of using the default hydrous ferric oxide as the sorbent were also presented in order to provide an indication of the adsorption range of Fe oxides.

5.4. Results

5.4.1. Geology and soil characteristics

The soils in Bwaise III slum consisted of 3 distinct layers: 1) a top layer containing loamy soils mixed with fill material; 2) a middle layer containing whitish/grey sands with small amounts of clay; and 3) a thick bottom layer containing stiff grey clay soils that extended to depths greater than 4 m below ground level (Fig. 5.3). Grain size analyses confirmed the distinction of the layers (Table 5.1) with soils in the middle layer having the highest portion of sand (66%). The bottom layer had the highest proportion of clay (38%). The water table was always encountered in the middle sandy layer implying that this layer formed the shallow alluvial aquifer whereas the upper and bottom layers acted as (semi) confining layers: confined at the bottom with a stiff clay layer and semi-confined at the top with loamy soils. Water always gushed up whenever the middle layer was struck during augering. The average thickness of this aquifer was 0.75 m (range of 0.5 -1.5 m) (Table 5.1, Fig. 5.3). Geo-available metals were low (less than 0.2% by soil weight, Table 5.1) implying that their availability to interact with nutrients in groundwater was low. The soil OC ranged from 0.6 to 0.8% by weight of soil (Table 5.1). The pH of the soils ranged from 5.2 to 6.8 indicating acidic to neutral soils.

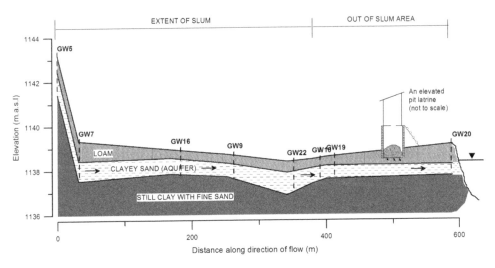

Figure 5.3:   Hydro-geologic cross-section along profile x-x of the shallow alluvial aquifer in Bwaise III parish slum. The clayey sand formed the shallow aquifer in the area. The bottom layer of clay was an impervious confining layer. The arrows represent the direction of groundwater flow.

Table 5.1:   Average soil properties (with standard deviations) of the different stratigraphic units in Bwaise III slum.

Soil layer	Depth (m)	pH (-)	OC (%)	Geo-available metals				CEC (meq/100g)	Grain size distribution		
				Fe (mg/kg)	Al (mg/kg)	Ca (mg/kg)			Sand (%)	Silt (%)	Clay (%)
Top	0 - 0.9	6.8	0.8	399	156	663		8.4	58	16	26
(n=20)		(±0.7)	(± 0.2)	(± 283)	(± 72)	(± 675)		(± 2.7)	(± 12)	(± 10)	(± 9)
Middle	0.5 - 2	6.7	0.7	373	125	481		5.8	66	13	21
(n=16)		(±0.8)	(± 0.2)	(± 159)	(± 67)	(± 413)		(± 2.4)	(± 12)	(± 6)	(± 8)
Bottom	> 2	5.2	0.6	308	74	140		6.4	50	12	38
(n=19)		(±1.0)	(± 0.2)	(± 187)	(± 41)	(± 89)		(± 1.9)	(± 9)	(± 4)	(± 9)

5.4.2. Groundwater flow

Water table contours indicated that groundwater consistently flowed from North East to South West throughout the sampling period (Fig. 5.4). Contours were more closely packed in the upstream areas than the downstream (Fig. 5.4) indicating groundwater inflow from the upper areas. We estimated a mean hydraulic gradient (i) of 0.02 m/m in the upper areas (influx zone) and 0.001 – 0.002 m/m in the lower areas where groundwater discharges. This gives a mean hydraulic gradient of 0.011. The water table was very shallow (average depth of 0.9 meters below ground level (m.b.g.l) in dry season and 0.5 m.b.g.l in the wet season) and fluctuated by about 0.5 m (Fig. 5.5). The hydraulic conductivity (K) of the shallow sandy aquifer ranged from 1 - 5 m/d with an average value of 2.24 m/d for 5 locations distributed throughout the study site (Table 5.2). This value is comparable with previous estimates in the same study area (Nyenje et al., 2013a; Kulabako et al., 2008).

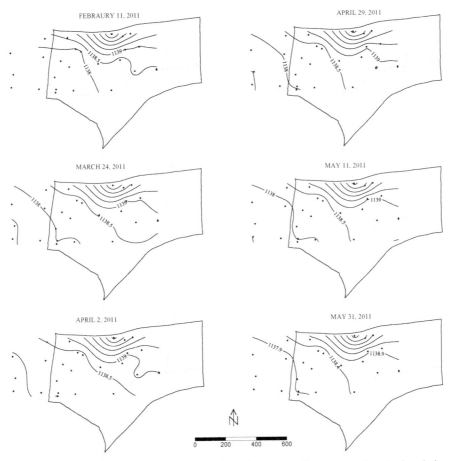

Figure 5.4:   Shallow groundwater contours based on manually measured water levels in Bwaise III slum from February to May 2011.

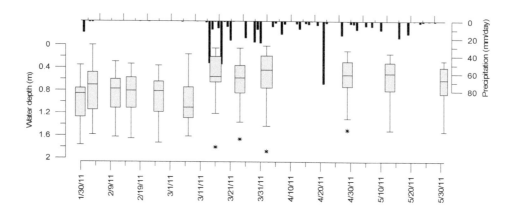

Figure 5.5: Variations in water table depth (m.b.g.l) and rainfall during the dry season
(Jan - March) and the wet season (starting from 15th March). Results are
presented as median daily variations in depths for all wells represented by box
whisker plots showing the minimum, maximum, median, lower quartile and
upper quartile. The star (∗) represents outliers.

Table 5.2:   Estimates of hydraulic conductivities (K, m/d) of the shallow alluvial aquifer
at different locations of the study area

No.	Location	Test 1	Test 2	Average
1	Near spring	2.86	1.83	2.35
2	GW16	1.37	0.62	1.00
3	GW1	4.92	5.18	5.05
5	GW4	0.61	1.50	1.06
6	GW10	1.23	2.29	1.76

We estimated the lateral average groundwater velocity using the Darcy equation. Using an
average hydraulic conductivity (K) of 2.24 m/d, an estimated porosity of 0.24 (Kulabako et
al., 2007, 2008) and the maximum hydraulic gradient of 0.02 (upper flow areas), the
maximum average groundwater flow velocity was 0.2 m/d. Using the minimum hydraulic
gradient of 0.002 in the lower flow areas, a much lower flow velocity of 0.02 m/d was
obtained. These velocities are similar to those estimated by Nyenje et al. (2013a) in the same
study area. Assuming that the residence time in the unsaturated zone is negligible, the
average flow distance from the upper flow boundary to the lower flow boundary was
estimated to be 800 m (see Fig. 5.1). Therefore, using the maximum estimated flow velocity
of 0.2 m/d, the residence time in the studied aquifer was about 11 years (= 800m/0.2 m/d).
Using the lowest estimated flow velocity of 0.02 m/d, the residence time was about 110
years. So the average residence time (average of maximum and minimum) in the study area
was about 60 years.

5.4.3. Hydrochemistry

Hydrochemistry and nutrients in shallow groundwater

Table 5.3 shows the descriptive statistics of the shallow groundwater beneath the slum area. The complete list of all hydrochemical data is presented in Table 5.4. The shallow groundwater was highly mineralized with high EC values (average of 1619 µS/cm; Table 5.3, Fig. 5.6) compared to the EC of inflowing groundwater (spring water; average 502 µS/cm; Table 5.3) and precipitation (30 µS/cm). Some wells (e.g. GW5, GW7, GW15 and GW17) had extremely high values of EC (up to 3000 µS/cm, Table 5.4). These wells were randomly spread within the slum area implying that a number of potential point pollution sources, such as pit latrines, existed. For example, wells GW5 and GW15 were located immediately downgradient of the pit latrines (personal observation).

Table 5.3:   Descriptive statistics of the hydrochemistry of the shallow groundwater beneath Bwaise slum. Values of variables are given in mmol/L except for EC (Electrical conductivity, µS/cm), pH (-) and T (temperature, °C).

			Shallow groundwater		
Parameter	Rainfall*	Spring	Within the slum area	Transect 1	Transect 2
EC	30	502	1619	1451	987
pH	6.1	4.8	7.2	6.68	6.58
T	24.2	25.9	26.5	26.6	26.3
Na^+	0.03	0.950	4.213	2.49	2.023
K^+	0.04	0.213	2.300	1.602	0.619
Mg^{2+}	0.01	0.123	0.800	0.963	0.479
Ca^{2+}	0.12	0.883	2.000	2.703	1.464
NH_4^+	0.05	0.190	2.230	1.843	1.071
Cl^-	0.08	1.391	3.176	2.286	1.254
HCO_3^-	0.07	0	10.78	10.82	8.895
SO_4^{2-}	0.02	0.032	0.364	0.662	0.084
NO_3^-	0.03	2.293	0.222	0.106	0.034
** PO_4^{3-}	0.004	0.0045	0.0068	0.0075	0.0079
Fe^{2+}	0	0.0029	0.316	0.619	0.800
Mn^{2+}	0.001	0.0072	0.0316	0.0366	0.054
DOC (as C)	-	0.41	1.07	1.68	1.93

*Data from Chapter 3 and Kulabako et al. (2007)

** Well GW2 was not considered when computing the average values for PO_4^{3-} because it had abnormally high concentrations of PO_4^{3-}.

The pH of the shallow groundwater was generally near neutral (pH = 6.6 - 7.2) with exception of the spring and some upgradient wells (e.g. LS and SW), which were acidic (pH < 5) since they were likely directly recharged by the more oxic and acidic groundwater flowing from the upper regolith and fractured aquifers (see Flynn et al., 2012; Nyenje et al., 2013b). The average Cl^- concentration was 3 mmol/L (Table 5.3; Fig. 5.6), with highs of 14 mmol/L also measured (e.g. GW15; Table 5.4). This average value is about 40 times more than the average Cl^- concentration in rainfall water (0.08 mmol/L; Table 5.3) and 2.3 times the average Cl^- concentration of inflowing groundwater (Spring: 1.38 mmol/L). This suggested that the shallow groundwater beneath the slum area was heavily contaminated and most likely from wastewater leaching from on-site sanitation systems in the slum area. Wastewater contains high amounts of organic matter, which when degraded, can lead to increased concentrations of nutrients (PO_4^{3-}, NH_4^+, NO_3^-) and ions such as Cl^-, Na^+, Ca^{2+}, K^+ and HCO_3^- in groundwater (Navarro and Carbonell, 2007; Ptacek, 1998). Consequently, the shallow groundwater beneath the slum area was characterized by high concentrations of carbon (1.1 - 1.9 mmol/L as DOC), nutrients (N and P) and cations, particularly Na^+ (up to 13.52 mmol/L) and Ca^{2+} (up to 6.13 mmol/L) and K^+ (up to 15 mmol/L) (Table 5.3; Table 5.4). NH_4^+ was the dominant form of nitrogen (average of 2.23 mmol/L; range of 1 - 3 mmol/L) (Fig. 5.6; Table 5.3). NO_3^- was generally absent (< 0.2 mmol/L), except in wells along the upper flow boundary and the spring, where high concentrations (average of 2.2 mmol/L, Table 5.3) were measured. The high NO_3^- concentrations in the upgradient wells were consistent with low pH observations, resulting from acid rain recharge in the catchment (Nyenje et al., 2013a) and the release of H+ following the conversion of NH_4^+ to NO_3^- by nitrification.

Phosphate (PO_4^{3-}) concentrations ranged from 1.3 - 13 μmol/L with an average value of 6 μmol/L (Table 5.3, Fig. 5.6). Well GW2 was an exception because the PO_4^{3-} concentrations were abnormally high (up to 200 μmol/L; Fig. 5.6). This well was installed near an organic waste dump subject to incineration with charcoal encountered in the unfiltered groundwater. It is likely that P transport here was affected by macro-pore flow. The dominant water types in the shallow aquifer were $NaHCO_3$ and $CaHCO_3$. In some instances $KHCO_3$ (upgradient wells) and NH_4HCO_3 (along Nsooba channel) water types were encountered (Fig. 7B). Ca^{2+} and Na^+ were the dominant cations.

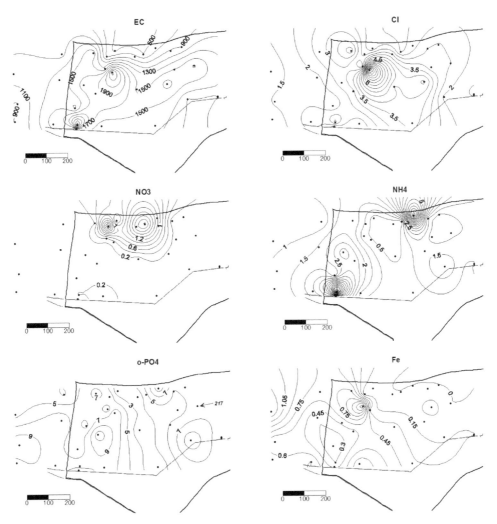

Figure 5.6   Evolution of average values of EC (µS/cm) and average concentrations of Cl⁻, NO_3^-, NH_4^+, o-PO_4 (or PO_4^{3-}) and Fe^{2+} (in mmol/L except PO_4^{3-} which is in µmol/L) in Bwaise III slum. Values were interpolated using Kriging techniques. Well GW5 was omitted from the interpolations of EC and Cl⁻ because it had abnormally high values. The blue lines represent the drainage network. The scale bar displays metres.

Chapter 5

Table 5.4: A complete hydro-chemical dataset of shallow groundwater in Bwaise III slum. Values of variables are given in mmol/L except for EC (Electrical conductivity, µS/cm), pH (-) and T (temperature, °C).

Well No.	Xcoord	Ycoord	Date	EC	pH	T	Na^+	K^+	Mg^{2+}	Ca^{2+}	NH_4^+	Cl^-	HCO_3^-	SO_4^{2-}	NO_3^-	Fe^{2+}	Mn^{2+}	PO_4^{3-}
GW1	451058	38664	3/21/2011	1164	7.0	24.1	1.18	1.15	0.50	1.84	1.64	1.28	11.78	0.13	0.09	0.20	0.04	0.004
GW1	451058	38664	3/28/2011	985	6.9	25.3	1.10	0.78	0.42	2.00	0.00	1.40	7.44	0.28	0.02	0.14	0.03	0.014
GW1	451058	38664	2/15/2011	1315	7.0	25.7	2.72	1.23	0.81	2.61	1.07	1.59	12.84	0.03	0.06	0.14	0.04	0.003
GW1	451058	38664	3/2/2011	1289	6.4	25.8	2.30	1.61	0.95	2.63	1.36	1.06	10.86	0.01	0.00	0.06	0.05	0.002
GW2	450946	38808	3/28/2011	1560	7.2	30.8	1.76	3.53	1.08	1.18	0.50	2.27	11.30	0.56	0.04	0.01	0.00	0.118
GW2	450946	38808	2/10/2011	1920	7.2	30.1	2.51	3.90	1.67	1.46	1.64	2.25	16.18	0.39	0.14	0.00	0.01	0.348
GW2	450946	38808	3/2/2011	1599	8.9	30.4	3.20	4.97	1.87	1.58	1.00	2.58	11.88	0.09	0.19	0.00	0.01	0.307
GW2	450946	38808	3/21/2011	1846	7.1	26	3.83	6.76	2.41	1.68	0.57	3.01	12.82	0.94	0.04	0.00	0.00	0.097
GW3	450889	38892	3/21/2011	977	7.7	26	2.39	1.06	0.33	1.03	0.57	2.61	6.26	0.27	0.01	0.00	0.01	0.005
GW3	450889	38892	3/2/2011	702	10.9	28.2	2.04	0.89	0.07	1.06	0.36	2.66	2.30	0.69	0.13	0.00	0.00	0.016
GW3	450889	38892	2/10/2011	727	10.2	26.1	0.75	0.42	0.09	0.98	2.02	1.36	2.38	0.81	0.12	0.01	0.00	0.003
GW3	450889	38892	4/7/2011	1220	8.0	27.8	5.03	2.68	0.76	2.28	0.57	3.59	8.58	0.25	0.04	0.00	0.02	0.001
GW4	450837	38875	2/10/2011	1040	7.0	25.5	1.54	0.60	0.28	1.16	3.92	3.43	5.50	0.68	0.16	0.02	0.01	0.011
GW4	450837	38875	3/2/2011	805	10.6	30.3	2.64	0.86	0.05	1.67	0.29	2.90	1.82	0.47	0.02	0.00	0.00	0.000
GW4	450837	38875	3/21/2011	932	7.1	26	3.13	1.01	0.49	1.92	0.00	1.23	5.06	0.12	0.01	0.05	0.04	0.004
GW5	450766	38890	3/21/2011	2240	7.1	27	3.58	3.74	0.68	1.62	2.57	4.85	7.18	1.05	2.44	0.00	0.01	0.016
GW5	450766	38890	3/2/2011	4970	7.9	28.4	11.24	14.88	1.22	1.22	18.56	7.29	32.47	0.87	0.91	0.00	0.00	0.007
GW5	450766	38890	2/10/2011	3810	7.4	24.9	10.35	11.34	2.13	2.51	6.00	8.39	16.04	1.31	0.33	0.00	0.00	0.008
GW6	450854	38790	2/10/2011	1461	7.0	25.1	3.37	1.63	0.56	2.10	0.57	3.36	10.22	0.04	0.26	0.14	0.02	0.003
GW6	450854	38790	3/21/2011	1567	6.9	24.2	5.71	2.80	1.03	2.43	1.00	2.24	11.06	0.05	0.00	0.41	0.03	0.001
GW6	450854	38790	5/17/2011	1498	7.1	23.9	12.49	1.50	0.48	1.16	0.64	2.39	10.38	0.07	0.02	0.21	0.02	0.007
GW7	450820	38724	5/17/2011	1697	7.2	25	3.25	2.65	0.47	2.32	0.79	3.45	12.60	0.23	0.02	0.06	0.03	0.003
GW7	450820	38724	2/10/2011	2060	7.1	26	8.99	1.87	1.29	1.54	2.14	4.87	15.12	0.00	0.36	0.17	0.02	0.006

Fate of sanitation-related nutrients in a shallow sandy aquifer

Well No.	Xcoord	Ycoord	Date	EC	pH	T	Na$^+$	K$^+$	Mg^{2+}	Ca^{2+}	NH$_4^+$	Cl$^-$	HCO$_3^-$	SO$_4^{2-}$	NO$_3^-$	Fe^{2+}	Mn^{2+}	PO$_4^{3-}$
GW7	450820	38724	3/2/2011	2020	8.6	27.4	6.30	3.39	0.97	3.02	2.28	5.62	12.70	0.06	0.04	0.00	0.03	0.003
GW7	450820	38724	3/21/2011	1402	7.0	24.7	5.03	2.45	0.64	2.66	0.64	4.76	6.38	0.66	0.10	0.00	0.01	0.008
GW8	450916	38652	3/2/2011	1121	7.6	27.5	1.76	0.56	0.29	1.03	2.86	1.96	7.92	0.02	0.11	0.03	0.01	0.006
GW8	450916	38652	3/28/2011	1166	6.9	24.6	1.89	0.99	0.44	1.81	1.71	2.70	8.38	0.35	0.04	0.20	0.03	0.010
GW8	450916	38652	2/15/2011	1108	7.0	28	2.99	1.28	0.35	0.97	2.50	2.16	8.90	0.01	0.17	0.06	0.02	0.016
GW8	450916	38652	5/16/2011	959	7.1	24.8	6.58	0.48	0.28	0.99	1.00	0.57	8.20	0.02	0.00	0.12	0.02	0.012
GW9	450528	38796	2/8/2011	1404	6.7	25.1	2.94	0.65	1.16	3.29	0.57	2.33	12.84	0.01	0.71	1.38	0.07	0.000
GW9	450528	38796	4/7/2011	1302	6.7	27.4	3.77	0.89	1.26	2.87	0.71	4.01	10.46	0.14	0.07	1.74	0.06	0.002
GW9	450528	38796	3/23/2011	1278	6.7	27.1	3.21	0.60	0.99	2.41	0.64	0.01	11.92	0.00	0.14	1.46	0.05	0.009
GW10	450429	38721	3/22/2011	1791	6.6	25.3	3.32	1.86	0.55	1.31	3.64	2.57	13.80	0.00	0.13	0.54	0.02	0.006
GW10	450429	38721	5/18/2011	1141	7.0	27.2	2.57	1.35	0.37	0.93	1.71	0.81	7.78	0.01	0.27	0.04	0.01	0.003
GW10	450429	38721	4/7/2011	2520	7.1	25.2	4.98	2.74	0.78	1.62	3.86	3.76	12.52	0.01	0.00	0.33	0.03	0.004
GW10	450429	38721	2/16/2011	1739	6.7	23.8	5.62	2.57	0.91	2.15	4.28	3.97	14.26	0.03	0.04	0.99	0.03	0.005
GW11	450388	38616	3/23/2011	1372	6.8	26.8	1.41	1.08	0.50	2.11	1.29	1.06	9.46	0.30	0.16	0.26	0.02	0.009
GW11	450388	38616	2/8/2011	1262	6.8	26.3	2.87	1.68	0.76	1.81	1.14	1.50	10.52	0.01	0.06	0.20	0.05	0.006
GW11	450388	38616	4/7/2011	1079	7.9	24.7	2.92	2.47	1.09	3.48	1.93	2.12	10.62	0.54	0.00	0.00	0.05	0.000
GW12	450511	38531	2/9/2011	1318	6.5	25.9	1.93	0.54	0.50	1.71	1.36	1.42	10.64	0.00	0.66	0.56	0.05	0.000
GW12	450511	38531	2/7/2011	1316	6.5	30.6	2.48	0.66	0.66	2.19	1.50	1.11	13.62	0.00	0.00	0.93	0.07	0.000
GW12	450511	38531	3/22/2011	1405	6.5	27.5	3.01	0.88	0.97	2.68	1.21	1.69	11.82	0.97	0.26	1.03	0.08	0.008
GW12	450511	38531	3/29/2011	1355	6.7	28.4	3.30	1.10	1.09	2.96	1.50	2.11	12.34	0.94	0.28	1.04	0.07	0.026
GW13	450762	38531	3/3/2011	1716	7.1	27.3	2.77	0.87	0.49	1.70	1.21	4.00	10.56	0.00	0.19	0.44	0.04	0.003
GW13	450762	38531	3/22/2011	1137	6.9	22	2.21	0.72	0.30	1.47	0.29	4.25	4.68	0.61	0.16	0.13	0.02	0.001
GW13	450762	38531	2/9/2011	1351	6.8	24	2.40	0.96	0.48	2.66	0.64	2.99	9.78	0.09	0.00	0.40	0.04	0.000
GW14	450474	38869	2/8/2011	1795	6.7	25.2	6.09	1.82	0.84	2.21	1.86	5.62	10.56	0.14	0.12	0.96	0.05	0.002

89

Well No.	Xcoord	Ycoord	Date	EC	pH	T	Na^+	K^+	Mg^{2+}	Ca^{2+}	NH_4^+	Cl^-	HCO_3^-	SO_4^{2-}	NO_3^-	Fe^{2+}	Mn^{2+}	PO_4^{3-}
GW14	450474	38869	4/7/2011	1425	6.9	26.1	4.94	2.48	0.65	1.38	1.93	2.88	8.48	0.03	0.04	0.05	0.02	0.005
GW14	450474	38869	3/3/2011	1738	6.8	31.6	5.96	1.65	0.91	2.31	2.21	4.67	8.20	0.02	0.36	1.28	0.05	0.016
GW15	450562	38780	3/22/2011	2290	6.7	25.3	9.56	2.20	0.87	2.40	0.50	9.02	6.94	2.16	0.07	0.10	0.03	0.002
GW15	450562	38780	3/29/2011	2770	6.9	26.2	13.52	3.04	1.33	3.08	0.50	13.54	10.44	1.96	0.08	0.34	0.05	0.018
GW15	450562	38780	5/18/2011	2830	6.9	26.1	10.98	2.28	1.10	2.65	0.71	7.05	11.92	0.73	0.76	0.28	0.04	0.006
GW16	450676	38701	3/3/2011	1370	7.1	27.1	4.31	1.58	1.06	2.65	0.43	3.97	8.74	0.14	0.18	0.26	0.07	0.002
GW16	450676	38701	2/9/2011	1551	6.6	26.1	5.92	1.70	1.32	2.51	0.21	3.98	10.04	0.16	0.00	0.47	0.09	0.001
GW17	450397	38539	3/23/2011	2650	7.1	27	4.84	3.92	0.69	2.35	7.50	4.63	17.78	0.33	0.12	0.21	0.04	0.002
GW17	450397	38539	5/18/2011	2940	7.5	26.2	5.64	4.38	0.75	2.06	10.35	2.45	23.09	0.04	0.14	0.05	0.03	0.003
GW17	450397	38539	3/3/2011	2980	7.2	25.9	5.13	4.21	0.81	2.77	12.14	4.99	21.01	0.00	0.25	0.50	0.04	0.000
GW17	450397	38539	2/15/2011	2920	7.1	25.9	6.37	4.66	0.95	2.83	9.28	0.94	23.33	0.00	0.36	0.40	0.05	0.027
GW21	450393	38519	3/23/2011	960	6.6	26.2	1.81	1.37	0.57	1.94	1.14	1.13	7.92	0.32	0.09	0.44	0.04	0.010
GW22	450484	38682	3/3/2011	1883	7.2	25.9	3.68	3.50	1.11	2.38	3.21	3.49	15.12	0.01	0.23	0.29	0.02	0.006
GW22	450484	38682	3/26/2011	1653	6.6	26	3.26	3.09	1.16	2.68	2.28	2.74	9.40	2.72	0.13	0.36	0.03	0.018
GW22	450484	38682	2/9/2011	1878	6.8	24.7	3.69	3.23	1.13	3.18	2.93	2.36	14.66	0.11	0.00	0.57	0.03	0.014
GW23	451058	38660	2/15/2011	856	6.6	27.3	3.05	0.89	0.73	2.21	0.36	1.54	7.64	0.34	0.05	0.41	0.03	0.005
spring	450708	38862	5/18/2011	533	5.0	25.8	0.78	0.17	0.07	0.65	0.14	0.78	0.00	0.04	2.94	0.00	0.00	0.002
Spring	450708	38862	2/9/2011	519	4.7	25.7	0.51	0.17	0.05	1.23	0.00	1.90	0.00	0.04	2.02	0.01	0.00	0.002
spring	450708	38862	3/26/2011	455	4.6	26.1	1.56	0.30	0.24	0.77	0.43	1.49	0.00	0.02	1.91	0.00	0.02	0.010
SW1	450538	38853	2/16/2011	658	5.7	26.8	1.72	0.32	0.23	0.32	0.91	2.26	1.06	0.30	2.19	0.01	0.01	0.007
SW2	450575	38878	3/4/2011	463	6.6	31.6	2.07	0.48	0.29	0.67	0.21	1.61	2.20	0.03	0.28	0.08	0.01	0.001
LS10	450668	38850	2/16/2011	772	3.5	25.7	3.18	0.50	0.44	0.73	0.14	3.06	0.00	0.37	1.94	0.01	0.03	0.001

Well No.	Xcoord	Ycoord	Date	EC	pH	T	Na$^+$	K$^+$	Mg^{2+}	Ca^{2+}	NH$_4^+$	Cl$^-$	HCO$_3^-$	SO$_4^{2-}$	NO$_3^-$	Fe^{2+}	Mn^{2+}	PO$_4^{3-}$
Transect 1																		
GW18	450315	38747	3/4/2011	1807	6.8	27.3	2.25	3.41	1.12	2.46	2.93	2.22	17.06	0.01	0.39	0.54	0.04	0.018
GW18	450315	38747	4/4/2011	866	6.9	23.3	1.78	1.89	0.59	1.70	0.50	1.75	7.62	0.09	0.04	0.07	0.04	0.004
GW18	450315	38747	2/7/2011	1505	6.9	28.4	2.54	3.48	1.19	3.13	1.21	1.62	12.62	0.16	0.19	0.28	0.07	0.004
GW19	450326	38865	3/4/2011	1019	6.5	27.1	1.51	0.54	0.34	0.93	0.71	1.59	6.70	0.05	0.16	0.31	0.02	0.003
GW19	450326	38865	3/23/2011	1044	6.3	26.5	3.33	1.01	0.75	1.55	0.50	3.87	6.92	0.05	0.00	1.04	0.05	0.003
GW20	450285	38542	3/4/2011	1126	6.6	25.6	1.69	0.82	0.60	1.84	2.71	1.86	11.16	0.00	0.10	0.80	0.02	0.008
GW20	450285	38542	2/7/2011	1450	6.7	29.3	2.59	1.49	1.09	2.65	2.57	2.34	12.00	0.00	0.00	0.11	0.03	0.010
GW20	450285	38542	5/19/2011	1657	6.9	25.8	2.29	1.20	1.10	2.69	2.50	1.21	13.12	0.15	0.04	0.71	0.03	0.004
GW20	450285	38542	4/4/2011	2270	6.8	26.7	4.15	1.49	1.89	6.13	2.50	4.33	10.72	4.84	0.06	0.69	0.04	0.003
GW20	450285	38542	2/8/2011	1403	6.6	25.4	2.03	0.97	0.70	2.91	1.86	2.13	10.46	0.00	0.14	0.90	0.03	0.004
GW20	450285	38542	3/23/2011	1815	6.6	26.9	3.20	1.31	1.21	3.74	2.28	2.23	10.62	1.93	0.04	1.36	0.04	0.022
Transect 2																		
Gw24	450103	38554	5/19/2011	846	6.6	25	1.21	0.57	0.54	1.62	0.86	0.95	7.82	0.10	0.05	0.52	0.09	0.014
GW24	450103	38554	4/4/2011	890	6.7	24.6	1.35	0.71	0.51	1.73	1.07	1.21	8.14	0.19	0.06	0.64	0.07	0.005
GW24	450103	38554	3/4/2011	980	6.5	26.5	1.80	0.82	0.64	2.17	1.57	1.08	9.44	0.02	0.00	0.87	0.08	0.006
GW25	450116	38637	4/4/2011	770	6.7	26.2	1.55	0.71	0.28	0.79	0.86	1.80	5.62	0.25	0.04	0.15	0.02	0.021
GW25	450116	38637	3/4/2011	1004	6.7	27	2.76	0.64	0.38	1.41	1.43	1.40	8.38	0.03	0.01	0.72	0.03	0.002
GW26	450098	38775	3/4/2011	1199	6.5	28.9	2.61	0.46	0.49	1.36	1.00	1.16	11.72	0.00	0.00	1.25	0.04	0.002
GW26	450098	38775	5/19/2011	1223	6.5	26.1	2.89	0.41	0.50	1.16	0.71	1.18	11.16	0.00	0.09	1.47	0.05	0.007

Evolution of water quality and nutrients along the flow path

Wells located along the transects downgradient of the slum area (e.g. GW 24, 25 and 26, about 200 m from the slum boundary) tended to have lower EC values (< 1200 µS/cm) than those within the slum area (Fig. 5.6). A few wells located along the upper flow boundary (e.g. SW, LS and GW3) also tended to exhibit lower EC values than those within the slum area. Hence, we observed a spatial evolution of EC values along the direction of groundwater flow. This was supported by chloride measurements whereby Cl^- concentrations of the downstream wells (G24, 25 and 26) did not exceed 1.8 mmol/L whereas those in Bwaise slum (Katoogo and St. Francis zones) ranged from 1.5 (GW 11) up to 9 mmol/L (GW 15). The observed evolution of EC and Cl^- is consistent with what would be expected when wastewater in the slum area infiltrates the aquifer. The observed water quality types along the direction of groundwater flow also showed similar variations (Fig. 5.7). For example, wells along upper flow boundary had NO_3^- as the dominant N species (Fig. 5.7) whereas further downgradient, NH_4^+ from wastewater leachate was the most dominant. Similar trends were also observed with the dominant cations (Ca^{2+} and Na^+). Ca^{2+} was the most dominant cation in the downgradient areas of the slum area (wells 20, 18, 19. 24, 25 and 26) whereas in the upper areas, Na type dominated water started to appear. The wells GW2 and GW5, located at the extreme upper flow boundary had K^+ as the dominant cation. A few wells near the Nsooba channel had NH_4^+ as the dominant cation, suggesting effects of inundation on groundwater from Nsooba channel. This channel is heavily polluted with high concentrations of NH_4^+ (0.5 - 3 mmol/L) and tends to flood during heavy storms (Nyenje et al., 2013a; Nyenje et al., 2014).

The spatial variations in nutrients were a result of both groundwater flow and a number of transformation processes. For example, the most dominant form of dissolved N in the shallow alluvial aquifer was NH_4^+. But as we moved upgradient, NO_3^- species progressively started to appear (Fig. 5.6). Phosphates did not show significant trends along the flow path. Concentrations were in most cases less than 10 µmol/L implying that P was regulated by various processes.

The redox environment

Fig. 5.7A shows the distribution of the most sensitive redox species; NO_3^-, Mn^{2+}, Fe^{2+} and SO_4^{2-}. With the exception of a few upgradient wells (e.g. SW1, SW2, LS10, spring, GW5), the concentrations of NO_3^- were relatively low (0 - 0.38 mmol/L) (Fig. 5.6, Table 5.4). On the other hand, high concentrations of Fe^{2+} (up to 1.74 mmol/L, Fig. 5.6, Fig. 5.7 and Table 5.4), and to a limited extent Mn^{2+} (0 - 0.09 mmol/L), were measured especially in the down gradient wells indicating that the shallow aquifer was generally anoxic. Consequently, NH_4^+ was also the dominant N species as mentioned above.

A close examination of NO_3^- variations (Fig. 5.6, Fig. 5.7B) showed that high NO_3^- concentrations occurred along the upper flow boundary (e.g. wells LS10, SW1, SW2, GW5 and the spring), which gradually reduced (from well GW 14) towards the downgradient areas in Katoogo zone and outside the slum area. This implied that denitrification took place along the direction of flow beneath the slum area. However, Fe^{2+} concentrations had the most remarkable results where a significant contrast existed between the upgradient wells and the downgradient wells. In most upgradient wells (e.g. SW1, SW2, LS10, GW5, GW2, GW3, GW4 and the spring), Fe concentrations were below 0.1 mmol/L (5 mg/L), whereas in most downgradient wells and a few upgradient wells (e.g. GW14), concentrations were generally higher than 0.2 mmol/L (11 mg/L) with values reaching as high as 1.74 mmol/L (94 mg/L) measured (Fig. 5.7B; Table 5.4). This suggested that the redox environment was predominantly Fe-reducing, which is consistent with field observations that most groundwater samples turned brown within 5 minutes after sampling. This change in color of samples was most likely caused by oxidation of Fe^{2+} to Fe^{3+} due to exposure of the samples to the atmosphere. In addition, we observed brown rusty deposits in the soil at the interface between the middle and bottom layer. These deposits were probably pockets of oxidised Fe^{3+}. Similar observations were made in previous studies (e.g. Nyenje et al., 2013a).

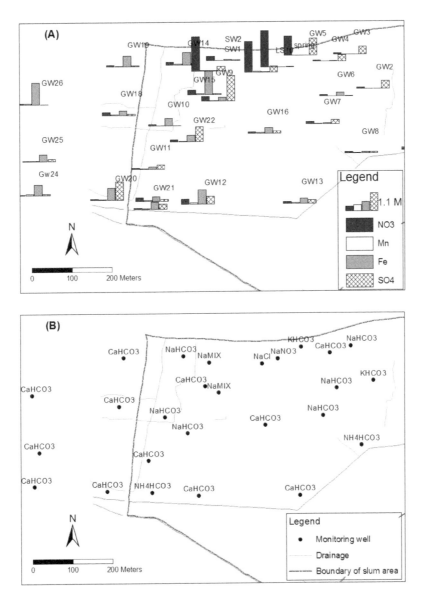

Figure 5.7:   (A) Distribution of redox species (means) and (B) classification of the most dominant water quality types in shallow groundwater in Bwaise slum.

A few wells (e.g. GW15, GW14, GW9, and GW7) had relatively high concentrations of both Fe^{2+} and NO_3^- indicating that these wells were located at the transition between nitrate-reducing to Fe/Mn-reducing environments. However, this occurrence could also have been caused by the mixed groundwater types during sampling given that our wells were fully penetrating.

SO_4^{2-} was generally present in high concentrations (range of 0 - 4.8 mmol/L, mean 0.36 mmol/L) indicating that conditions were not yet sulphate-reducing (Fig. 5.7, Table 5.4). This is also consistent with our field observations of the absence of the H_2S smell. In the last transect (wells GW24, GW25 and GW26), however, groundwater seemed to be depleted of SO_4^{2-} (Fig. 5.7) suggesting that sulphate-reducing conditions were starting to occur.

Although our results indicated that the redox conditions in the alluvial aquifer were Fe-reducing, the concentrations of Mn^{2+} were relatively low compared to Fe^{2+}, even though Mn-oxides are reduced at a higher redox potential than Fe oxides. The limited presence of Mn^{2+} in these rather strongly reducing conditions (Fe-reducing) suggests that Mn minerals were less abundant in the aquifer soils. It is also likely that Mn^{2+} ions were regulated by the solubility of Mn-minerals like rhodochrosite ($MnCO_3$) since HCO_3^- was a dominant ion in the shallow aquifer. Rhodochrosite has been shown in many studies to act as a sink of Mn^{2+} in anoxic aquifers (Jensen et al., 2002; Massmann et al., 2004; Matsunaga et al., 1993).

5.4.4. Geochemical modeling results

Mineral saturation indices

The relevant mineral saturation indices (SIs) calculated using the PHREEQC code are presented in Fig. 5.8. Samples N18, N35 and N34 (Table 5.4) were excluded from SI calculations because their pH measurements were considered to be erroneous. One of the most interesting observations was the very high partial pressure of CO_2 (pCO_2), which generally ranged from $10^{(-1.0)}$ to $10^{(-0.6)}$ atm. This is about 300 - 1000 times the atmospheric pCO_2 gas pressure of $10^{(-3.5)}$. Given that Bwaise slum was a reclaimed wetland, both the decaying organic matter related to wetland vegetation (0.6 - 0.8% as soil OC) and the large input of organic matter via wastewater leaching from pit latrines (up to 1.9 mmol/L as DOC), can explain the observed high CO_2 pressures. As noted in literature, it was evident in our study that increases in pCO_2 corresponded with a decrease in the SI values of calcite and the subsequent decrease in pH values (Fig. 5.8, Fig. 5.9). On the other hand, the decrease in the pCO_2 resulted in an increase in pH and subsequent increase of SI of calcite (Fig. 5.9). Indeed, some of our samples were supersaturated with respect to calcite at low values of pCO_2 (Fig. 5.8). Hence, the CO_2-partial pressures likely played an important role in regulating calcite in groundwater.

Groundwater was generally near saturation with respect to calcite and rhodochrosite (-0.6 < SI < 0.7). Most samples were, however, under-saturated with respect to hydroxyapatite (Ca_5(PO_4)$_3$OH) (SI < -2), suggesting that this mineral was not present and did not regulate PO_4^{3-} concentrations. If this mineral phase was present, it is unlikely that it would have remained undersaturated because the groundwater residence time was long (60 years). The well GW2 was an exception because it had a high SI value of hydroxyapatite (> 2; Fig. 5.8) owing to the abnormally high PO_4^{3-} concentrations. This pointed to localized geochemical conditions likely caused by contamination via macro pores.

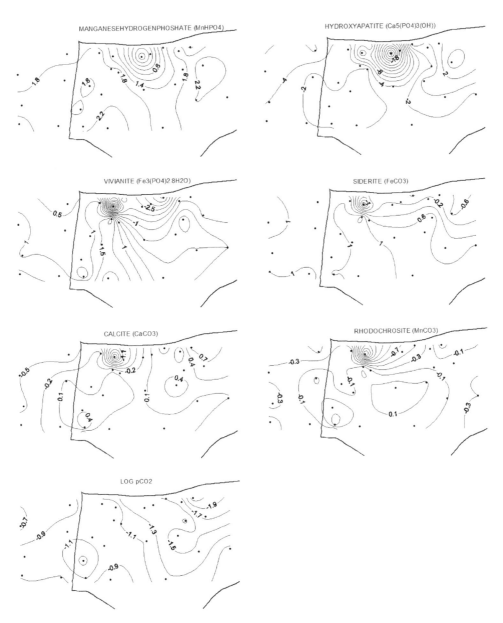

Figure 5.8:   Distribution of mean saturation indices of minerals likely to regulate fate of PO_4^{3-} and the mean partial pressures of CO_2 (pCO_2).

A number of samples were slightly supersaturated with respect to siderite ($0.6 < SI < 1$) especially in the more reduced zones where Fe^{2+} concentrations were high. This suggested that this carbonate phase did not control the concentrations of Fe^{2+} in those areas. In the upgradient less reduced areas, groundwater was undersaturated with respect to siderite ($FeCO_3$), which is consistent with the observed low concentrations of Fe in these areas (Fig. 5.6, Fig. 5.7). Most groundwater samples were consistently supersaturated with respect to vivianite ($SI > 1$) suggesting that this mineral did not regulate PO_4^{3-} concentrations. Mobile Fe^{2+} was likely supplied through reductive dissolution of Mn and Fe-oxides and hydroxides. Groundwater was also consistently oversaturated with respect to $MnHPO_4$ with SI values ranging from 1.8 - 2.2 except in the upper flow areas where SI values ranged from -0.6 and 0.6. This suggests that $MnHPO_4$ could have regulated PO_4^{3-} concentrations only in the upgradient parts of the aquifer.

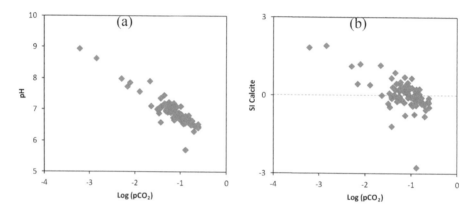

Figure 5.9:   The pH and samples and SI of calcite plotted against the partial pressure of CO_2 (pCO_2).

PHREEQC results of phosphorus sorption

The PHREEQC model code was used to simulate the sorption of PO_4^{3-} to the Fe-oxides present in the aquifer as goethite. A plot of the modeled amount of adsorbed P and the equilibrium PO_4^{3-} concentrations in the aquifer is shown in Fig. 5.10. The figure shows that the amount of goethite present would get saturated at equilibrium PO_4^{3-} concentrations of about 8×10^{-6} μmol/L, with a maximum P sorption capacity of 2.1 mmol/Kg. Here, saturation refers to a state where there is a rapid increase in the equilibrium ortho-phosphate concentrations while the amount of P sorbed remains relatively constant. The average concentration of PO_4^{3-} measured in the aquifer was 6 μmol/L and this value is much higher than the concentrations that would be present (8×10^{-6} μmol/L) if sorption to Fe-oxides controlled P concentrations in the aquifer. This implied the sorption sites on the Fe-oxides present in the aquifer as goethite ($\cong$ 400 mg/Kg: Table 5.1) were already exhausted (or saturated) and that the adsorption of P to Fe-oxides did not play a significant role in regulating P concentrations in the aquifer. In case ferrihydrite was the dominant Fe-oxide present, the results (in Fig. 5.10) show that the amount of P adsorbed would be 10 times more (21 mmol/Kg) but the equilibrium PO_4^{3-} concentrations would remain the same.

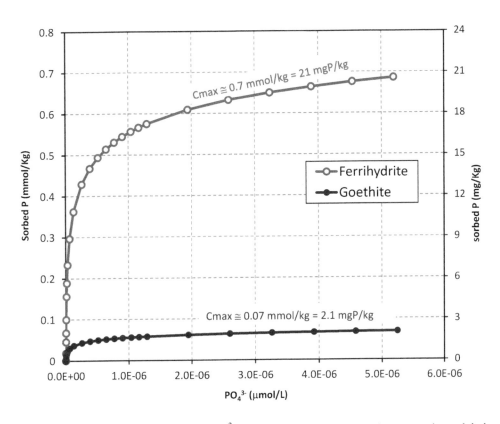

Figure 5.10:   A plot of dissolved PO_4^{3-} concentrations in groundwater and modeled phosphorus concentrations adsorbed on the Fe-oxides in the aquifer material.

5.5. Discussion

5.5.1. Evidence of pollution from on-site sanitation

The shallow alluvial sandy aquifer (about 1 m thick) was highly permeable with a mean hydraulic conductivity of 2.24 m/day, which suggests that it was prone to contamination. It was confined at the bottom by a stiff clay layer implying that contaminants spread mostly laterally in the direction of groundwater movement. Groundwater consistently flowed in the direction NE - SW with a calculated mean residence time of about 60 years.

Despite the presence of a large number of point pollution sources (essentially pit latrines - average number of 10 per hectare), which would otherwise suggest that diffuse pollution was taking place, our field findings showed that a contaminant plume existed. This plume was characterized by a front of polluted groundwater moving downstream in the direction of groundwater flow. Our results showed that the plume originated from wastewater infiltrating from on-site sanitation systems, particularly pit latrines (the dominant form of excreta disposal - see section 5.2). This is because we observed an evolution of EC values and Cl^- concentrations along the direction of groundwater flow. The breakthrough of EC peaks have for a long time been used to characterize contaminant plumes from wastewater loadings (e.g. Robertson, 2008; Robertson et al., 2012). The chloride ion is also a good conservative tracer of wastewater leaching from pit latrines (Graham and Polizzotto, 2013). The shallow groundwater in the slum area was characterized by high values of EC (up to 3000 μS/cm) and high concentrations of Cl^- (up to 9 mmol/L), which decreased along the direction of flow. The lowest concentrations (EC < 1200 μS/cm, Cl^- < 1.5 mmol/L) occurred in the last transect of monitoring wells (GW 24, GW 25 and GW 26) located near the swamp (Fig. 5.6). This transect, therefore represented relatively unpolluted shallow groundwater. More to this, the concentrations of ions (e.g. Ca^{2+}, Na^+, Cl^-, HCO_3^- and PO_4^{3-}) in the shallow groundwater beneath the contaminating slum area were always much higher than corresponding concentrations in the inflowing groundwater (spring) and precipitation (Table 5.3). The contaminant plume was also dominated by HCO_3-type water (up to 30 mmol HCO_3/L) indicating that the shallow aquifer received high organic loads from infiltrating wastewater. Organic matter is considered to be the principal source of HCO_3^- in non-calcareous soils (Appelo and Postma, 2007).

5.5.2. Presence of redox zones

Two distinct redox zones and a transition zone were evident in the contaminant plume. In the upper flow zones, groundwater contained elevated concentrations of NO_3^- (and Mn^{2+} to some extent) whereas Fe^{2+} was largely absent. In the lower flow zones in Bwaise slum, Fe (and some Mn^{2+} ions) was dominant whereas NO_3^- was absent. Hence, the shallow groundwater was NO_3-/Mn-reducing in the upper flow areas and Fe-reducing in the lower flow areas. The lower flow areas included large parts of Bwaise slum and areas downgradient of the slum area. The strong Fe-reducing conditions were reflected by the high concentrations of dissolved Fe^{2+} (Fig. 5.6). Between the upper and lower flow zones, groundwater contained all the three redox species Fe^{2+}, Mn^{2+} and NO_3^- suggesting the redox state of groundwater here was in transition (Mn-reducing). These changes in redox zones can be explained as follows: When partly oxidised nitrate-rich groundwater flowing from the upper regolith aquifers enters Bwaise III slum (see Nyenje et al., 2013b), it mixes with wastewater leachates from on-site sanitation. This will deplete oxygen during the degradation of the high organic carbon in the aquifer, hence resulting in anoxic conditions that are responsible for the observed decline in nitrate concentrations (or denitrification). We think that the source of carbon responsible for depleting oxygen and denitrifying nitrate originated from both the high DOC in wastewater leaching into the aquifer (1 - 1.9 mmol/L C; Table 5.3) and the soil organic carbon present in the wetland soils (0.6 to 0.8%; Table 5.1). DOC concentrations more than 0.4 mmol/L C are usually considered sufficient to denitrify nitrate (Rivett et al., 2008). Likewise, soil organic carbon contents in the range of 0.08 to 0.16 % are reported to be adequate to denitrify large quantities of nitrate (Hiscock et al., 1991; Trudell et al., 1986). Following further degradation of organic carbon from DOC and the soil, the aquifer becomes more reducing and in our study, it was Fe-reducing. In between the nitrate-reducing and Fe-reducing zones, a Mn-reducing zone exists, which in this study was evident by the presence of Mn^{2+} ions. Fig. 5.11 therefore shows a schematic of the nutrient transport processes and the redox zones observed in the studied shallow aquifer.

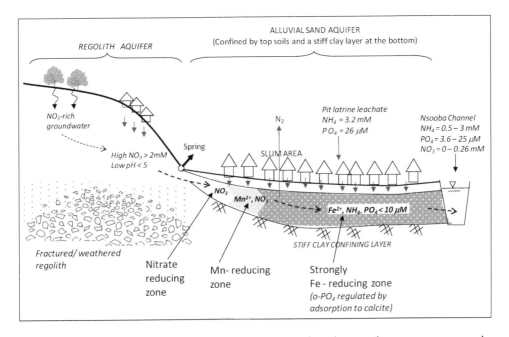

Figure 5.11:   A schematic of the redox zones, the fate of nutrients and transport processes in the shallow aquifer beneath the slum area (modified from Nyenje et al. (2013b). Not to scale.

5.5.3. Fate of nutrients (N and P)

The fate of nitrogen in the studied aquifer seemed to be largely controlled by changes in redox zones. In the upper flow boundary (which received nitrate-rich groundwater from the regolith aquifer), NO_3^- was dominant while NH_4^+ was largely absent. As groundwater flowed into Bwaise III slum and other downgradient areas, NH_4^+ became the dominant N species whereas NO_3^- disappeared. This suggested that NO_3^- was removed by denitrification (NO_3^- converted to free N_2 gas) as groundwater became more reducing (Fe-reducing) whereas NH_4^+ was likely added following wastewater infiltration from on-site sanitation systems in the slum area. Our findings are consistent with earlier studies in shallow alluvial aquifers where denitrification is believed to be an important process responsible for the removal of nitrate from groundwater originating from upland areas (e.g. Hoffmann et al., 2006; Rivett et al., 2008; Stigter et al., 2011). NH_4^+ concentrations ranged from $1 - 3$ mmol/L (average of 2.23 mmol/L). These concentrations were, slightly lower than the potential input from pit latrines of 3.2 mmol/L (57 mg/L; Nyenje et al., 2013a), which indicated that a 30% partial removal of NH_4^+ also took place. This removal, however, cannot be explained by cation exchange because of the low CEC of the sandy aquifer (< 6 meq/100g; Table 5.1). The partial removal of NH_4^+ was therefore likely due to ANaerobic AMMonium OXidation (ANAMMOX) as demonstrated in a related study by Robertson et al. (2012). Anammox is a bacteria-facilitated biological process that converts NH_4^+ directly to N_2 gas under anaerobic conditions. The presence of the anammox bacteria can be detected by extracting the DNA of microbial communities using the quantitative polymerase chain reaction (qPCR). Evidence for the presence of anammox bacteria in this study was, however, not carried out. In summary, the shallow sandy aquifer contributed to an almost complete removal of NO_3^- originating from the nitrate-rich inflowing groundwater due to denitrification, and to a 30% partial removal of NH_4^+ ($1 - 3$ mmol/L was left) originating from on-site sanitation (particularly, pit latrines), which was likely due to anaerobic oxidation. The concentrations of NO_3^- and NH_4^+ in Nsooba drainage channel where groundwater exfiltrates ranged from 0 to 0.26 mmol/L and $0.5 - 3$ mmol/L respectively (see Nyenje et al., 2013b, 2014). These concentrations are indeed close to those measured in the shallow groundwater in this study.

The fate of PO_4^{3-}, however, seemed to be governed by a range of complex processes because its concentrations were generally low and did not follow a particular trend along the groundwater flow path (Fig. 5.6). Measured PO_4^{3-} concentrations ranged from 1.3 to 13 μmol/L with a mean of 6 μmol/L. These values were very low compared to the potential input from pit latrines (2.4 mg/L or 26 μmol/L; Nyenje et al. 2013a). This implies that at least 75% of P originally present in wastewater effluents was removed within short distance during transport from the source of pollution to the aquifer. Previous studies show that most P is retained either in the vadose zone immediately below the source of pollution or in the aquifer material beneath the pollution source (Nyenje et al., 2013a; Robertson et al., 1998; Zanini et al., 1998; Zurawsky et al., 2004). Despite the high PO_4^{3-} removal in the studied aquifer, the concentrations measured (average of 6 μmol/L) were still slightly higher than the minimum required to cause eutrophication (0.075 mg TP/L or 2.4 μmol/L; Dodds et al, 1998). This groundwater can therefore still impair surface water upon groundwater exfiltration. The concentrations of PO_4^{3-} in Nsooba channel where groundwater discharged were, however, much higher ($3.4 - 25.3$ μmol/L; Nyenje et al., 2014) than those found in groundwater. This was probably due to additional P loads from grey water effluents to surface water as noted by Nyenje et al. (2014).

The attenuation of P following wastewater loading has for a long time been attributed to the strong adsorption affinity of P to metal oxides and calcite, and to mineral precipitation reactions with Fe, Mn, Ca and Al (Golterman, 1995; Reddy et al., 1999; Zanini et al., 1998). In this study, we used geochemical modeling to investigate the relative importance of these two processes in regulating P concentrations in the shallow aquifer. Results from the PHREEQC model indicated that mineral precipitation reactions played an important role in limiting PO_4^{3-} transport in groundwater. Normally, minerals whose SI values are near zero or near saturation ($SI = 0 \pm 0.5$) are considered capable of controlling mineral compositions in groundwater (Deutsch, 1997). Along the upper flow boundary of the studied slum area, conditions were slightly oxic and groundwater was near saturation with respect to $MnHPO_4$ and calcite implying that these mineral phases controlled the presence of PO_4^{3-} via precipitation (for $MnHPO_4$) and co-precipitation (for calcite). In the more reducing downgradient areas (the slum area), groundwater was near saturation with respect to rhodochrosite ($MnCO_3$) and calcite ($CaCO_3$) suggesting that these minerals regulated P concentrations by co-precipitation (Fig. 5.8). Calcite is widely reported to control PO_4^{3-} concentrations in contaminated groundwater (e.g. Cao et al., 2007; Golterman and Meyer, 1985; Griffioen, 2006; Richardson and Vaithiyanathan, 1995; Spiteri et al., 2007). Calcite was likely regulated by the high partial pressures of CO_2 (10^{-1} to $10^{-0.6}$) due to the high organic content in the aquifer. Mn carbonates, on the other hand, also have a high potential to precipitate because of their low solubility (Boyle and Lindsay, 1986). Groundwater in our study was supersaturated with respect to vivianite, siderite and $MnHPO_4$ and undersaturated (SI < -2) with respect to hydroxyapatite implying that these minerals were either not present or they were not reactive. In other studies, however, vivianite and hydroxyapatite are widely reported to regulate P in wastewater plumes from on-site sanitation (Golterman and Meyer, 1985; Ptacek, 1998; Spiteri et al., 2007; Zanini et al., 1998). There are a number of possible reasons why these minerals, were not reactive in this study. First, groundwater was strongly reducing (Fe-reducing) implying that there was continuous enrichment of Fe^{2+} and Mn^{2+}, which probably resulted in supersaturation levels of vivianite and siderite (due to reduction of Fe III oxides to Fe^{2+}) and $MnHPO_4$ (due to reduction of Mn IV oxides). Second, the high SI values of $MnHPO_4$ were likely maintained by the carbonate phase of Mn (rhodochrosite), whose SI values were always near saturation (see Fig. 5.8). Third, the concentrations of PO_4^{3-} in groundwater were low.

Apart of mineral precipitation reactions, sorption to the aquifer material could also have contributed to the high P attenuation observed in this study (> 75%). The adsorption of PO_4^{3-} is largely attributed to presence of Fe-oxides, although other sorbents like Al and Mn oxides and calcite are also considered important (Golterman, 1995). We used geochemical modeling to investigate the relative importance of the available Fe-oxides (about 400 mg/kg as goethite, Table 5.1) in regulating the transport of PO_4^{3-} by adsorption. Results from the PHREEQC model simulations suggested that if sorption to Fe-oxides was an important process, then equilibrium PO_4^{3-} concentrations in the aquifer would have been 8×10^{-6} µmol/L (Fig. 5.10). However, the measured average equilibrium PO_4^{3-} concentration in the aquifer was 6 µmol/L. This suggested that the Fe-oxide sorption sites present in the aquifer were few to cause a substantial decrease in P concentrations by adsorption. The shallow groundwater was largely anoxic (Fe-reducing) implying that reductive dissolution of Fe oxides could have depleted the adsorption capacity of the aquifer. Fig. 5.10 also shows that the P sorption capacity of the aquifer (C_{max}) was about 0.07 mmol/Kg (2.1 mg P/Kg). This value is much lower than the amount of available P in the sediments (18 - 50 mg/Kg; Nyenje et al., 2013a and Kulabako et al., 2008). These findings suggest that PO_4^{3-} could have been regulated by the adsorption of P to calcite or even clay minerals as suggested by de Mesquita Filho and Torrent (1993). But we think that calcite played an important role here as explained in the following text. Kulabako et al. (2008) investigated the adsorption capacity of the aquifer in the same study area by performing sorption and fraction experiments on a number of soil samples. Results from Kulabako's study showed that the aquifer, on the contrary, had a strong sorption capacity with a maximum sorbed phosphorus potential (C_{max}) of 600 mg/Kg and a Langmuir constant of 563 L/g. This study further performed multi-variate statistical correlations and found out that the Langmuir C_{max} was more correlated with Ca, pH, organic carbon and available P ($r^2 = 0.8$). pH is already known to affect P sorption whereby sorption increases with decreasing pH (Reddy et al., 1999) while organic carbon could have been related to presence of organic P. Although Kulabako's study did not determine Al, P adsorption to Al-oxides is generally reported to be greatly inhibited in presence of high organic matter (up to 1.9 mmol/L as DOC in this study) (Borggaard et al., 1990; Janardhanan and Daroub, 2010). We therefore conclude from these findings that the P sorption capacity of the aquifer sediment was related to Ca content and that sorption onto calcite could be contributing the high P retention in the studied aquifer. This conclusion is supported by the high Ca^{2+} and HCO_3^- concentrations we measured in the shallow groundwater (Table 5.3, Table 5.4) and the high Ca content in the sediment (Table 5.1). Groundwater was also near saturation with respect to calcite suggesting that calcite was present. Unfortunately, it is not possible to quantify the amount of calcite or carbonate content in the aquifer because it was not determined. Likewise, few mineralogical studies have been carried out in the study area.

In general, our findings suggest that the removal of nutrients in the lateritic alluvial sandy aquifers underlying the studied slum area was significant due to strong Fe-reducing conditions associated with long residence times. These conditions favored N removal by denitrification (for NO_3^-) and anaerobic oxidation (for NH_4^+), and PO_4^{3-} removal by co-precipitation with Ca and Mn carbonates - a natural attenuation process that may be preferred in urban poor areas without an end-of-pipe treatment. Hence, shallow alluvial aquifers with long residence times provide an important sink of nutrients even in heavily contaminated environments like unsewered urban slum areas.

5.6. Conclusions

We attempted to understand the processes governing the fate of sanitation-related nutrients (NO_3^-, NH_4^+ and PO_4^{3-}) in a shallow sandy aquifer underlying a poorly sanitized urban slum area in Kampala Uganda. Below are the most important conclusions from our study:

1. The shallow sandy aquifer in the slum area was heavily contaminated as indicated by the high values of EC and high concentrations of Cl^-, Na^+, Ca^{2+}, HCO_3^-, DOC and nutrients. Nutrients were predominantly present as NH_4^+ ($1 - 3$ mmol/L, average of 2.23 mmol/L) whereas the concentrations of PO_4^{3-} (average of 6 μmol/L) and NO_3^- (< 0.2 mmol/L) were low. We attributed this contamination to wastewater infiltrating from pit latrines.

2. Field data revealed the presence of a contaminant plume, which was characterized by decreasing values of EC and [Cl^-] downgradient of the groundwater flow line.

3. Distinct redox zones were identified along the groundwater path (NE-SW), whereby the groundwater in the upper flow zones was NO_3/Mn-reducing whereas within the slum area and further downstream, it was strongly Fe-reducing. Hence, groundwater was anaerobic and we attributed this condition to the long residence time (60 years) and the high organic loading (up to 1.9 mmol/L as DOC) originating from wastewater infiltration.

4. Variations in redox zones seemed to affect the mobility of sanitation-derived nutrients whereby nitrogen species were significantly removed from the aquifer. The high concentrations of NO_3^- (2.2 mmol/L) flowing from the upper regolith areas was almost completely removed by denitrification (< 0.2 mmol/L of NO_3 was left) in the alluvial aquifer beneath the slum area. Ammonium leaching directly into the aquifer from the slum area was also partially removed (by about 30%) because the measured concentrations were less than the potential input from pit latrines. Because the cation exchange capacity of the aquifer material was low (< 6 meq/100g), we concluded that anaerobic ammonium oxidation *(Anammox)* was responsible for the partial removal of NH_4^+, which was about 30%.

5. The concentrations of PO_4^{3-} in the shallow aquifer below the slum area were generally low (average of 6 µmol/L) compared to the potential input from pit latrines (26 µmol/L). Our findings showed that P was strongly attenuated by two processes: the adsorption of P to calcite in the aquifer and the co-precipitation of P with calcite and rhodochrosite. The adsorption of P to Fe-oxides present in the aquifer was found to be low because of the low Fe content (400 mg/Kg) in the aquifer sediments.

6. Our findings suggest that heavily contaminated groundwater discharging from poorly sanitized slum environments built on low-lying lateritic alluvial sandy aquifers may not be of major concern to eutrophication when groundwater exfiltrates to surface water systems. This is because these aquifer systems are highly anaerobic, due to long groundwater residence times, and can therefore effectively attenuate nutrients by precipitation and adsorption with mineral carbonates for phosphorus, and by denitrification and anaerobic oxidation which converts NO_3^- and NH_4^+ to free nitrogen.

Chapter 6

Phosphorus transport and retention in a channel draining an urban tropical catchment with informal settlements

Abstract

Urban catchments in sub-Saharan Africa (SSA) are increasing becoming a major source of phosphorus (P) to downstream ecosystems. This is primarily due to large inputs of untreated wastewater to urban drainage channels, especially in informal settlements (or slums). However, the processes governing the fate of P in these channels are largely unknown. In this study, these processes are investigated. During high runoff events and a period of base flow, we collected hourly water samples (over 24 hours) from a primary channel draining a 28 km^2 slum-dominated catchment in Kampala, Uganda and from a tertiary channel draining one of the contributing slum areas (0.54 km^2). The samples were analyzed for ortho-phosphate (PO$_4$-P), particulate P (PP), total P (TP), suspended solids (SS) and hydrochemistry. We also collected channel bed and suspended sediments to determine their geo-available metals, sorption characteristics and the dominant phosphorus forms. Our results showed that the catchment exported high fluxes of P (0.3 kg km^2 d^{-1} for PO$_4$-P and 0.95 for TP), which were several orders of magnitude higher than values normally reported in literature. A large proportion of P exported was particulate (56% of TP) and we inferred that most of it was retained along the channel bed. The retained sediment P was predominantly inorganic (> 63% of total sediment P) and consisted of mostly Ca and Fe-bound P, which were present in almost equal proportions. Ca-bound sediment P was attributed to the adsorption of P to calcite because surface water was near saturation with respect to calcite in all the events sampled. Fe-bound sediment P was attributed to the adsorption of P to Fe-oxides in suspended sediment during runoff events given that surface water was undersaturated with respect to Fe-phosphates. We also found that the bed sediments were P-saturated and showed a tendency to release P by mineralization and desorption. During rain events, there was a flushing of PP which we attributed to the resuspension of P-rich bed sediment that accumulated in the channel during low flows. However, first-flush effects were not observed. Our findings provide useful insights into the processes governing the fate and transport of P in urban slum catchments in SSA.

This chapter is based on: Nyenje, P.M., Meijer, L.M.G., Foppen, J.W., Kulabako, R., and Uhlenbrook, S., 2014, Phosphorus transport and retention in a channel draining an urban, tropical catchment with informal settlements: Hydrology and Earth System Sciences, v. 18, p. 1009-1025.

6.1.   Introduction

Phosphorus (P) derived from urban catchments in sub-Saharan Africa (SSA) is increasingly becoming a major cause of eutrophication of urban fresh water bodies (Nhapi et al., 2002; Nyenje et al., 2010). This is primarily attributed to the increasing release of untreated or partially treated wastewater into the environment especially in the informal settlements or slums. The number of informal settlements in most cities in SSA is growing rapidly following rapid urbanization and population growth. Furthermore, these areas often lack sewerage systems for collecting and treating wastewater while at the same time the existing on-site sanitation systems are usually poor. Consequently, most wastewater generated from these types of catchments is discharged untreated or partially treated into urban streams/channels resulting in the introduction of high concentrations of nutrients (nitrogen, N, and phosphorus, P) to downstream fresh water bodies (Bere, 2007; Foppen and Kansiime, 2009; Isunju et al., 2011; Katukiza et al., 2010b; Kulabako et al., 2010; Nhapi et al., 2002; Nhapi and Tirivarombo, 2004; NWSC, 2008). This has led to deterioration in the water quality of most urban fresh water bodies in SSA due to eutrophication. Phosphorus is considered to be the limiting nutrient for eutrophication (Reddy et al., 1999). However, current literature shows that there is limited research on P transport in urban catchments dominated with informal settlements especially in SSA.

A large number of studies in recent years have focused on understanding the dynamics of P transport during high and base flow periods (e.g. Blanco et al., 2010; Jordan et al., 2005; Peters and Donohue, 2001; Stutter et al., 2008; Zhang et al., 2007). It is often realized that high flows exhibit higher P concentrations predominantly in particulate form compared to base flows (or low flows). This is largely attributed to the flushing of pollutants and sediments from the catchment by the increased flow. The first-flush effect is also usually reported and it occurs when the first part of the storm runoff has substantially higher concentrations of pollutants than later parts (Deletic, 1998). A view therefore exists that during the rising limb of the hydrograph, there is an initial flushing of P-rich sediments generated from either terrestrial catchment runoff or from the resuspension of channel bed sediments (e.g. Blanco et al., 2010; Evans et al., 2004; Rodríguez-Blanco et al., 2013; Stutter et al., 2008; Zhang et al., 2007). First-flush effects are, however, not widely reported for dissolved nutrients such as NO_3^- and PO_4^{3-} because other mechanisms such as dilution and discharge from base flows tend to be dominant (Blanco et al., 2010; Chua et al., 2009; Evans et al., 2004; Jordan et al., 2005; Zhang et al., 2007).

There is also a large number of studies focusing on understanding the chemical processes, that influence the transport and fate of phosphorus in streams and rivers (Bedore et al., 2008; Evans et al., 2004; Froelich, 1988; Olli et al., 2009). These processes generally include precipitation and dissolution, adsorption to soils and sediments and redox reactions. The precipitation of P minerals usually occurs with Fe^{2+}, Al^{3+}, Ca^{2+} and Mn^{2+} ions and this normally leads to retention of P in sediments (Bedore et al., 2008; Evans et al., 2004; Golterman and Meyer, 1985; Reddy et al., 1999). In hard and alkaline fresh waters, most P is often retained by precipitating as hydroxyapatite (Golterman, 1995; Olli et al., 2009). The adsorption of P to Fe, Al or Mn oxides and hydroxides in sediments or soils is also another important process that contributes to the retention of P to bed sediments in river systems (Froelich, 1988; Golterman, 1995). In Ca-rich waters, P is also widely reported to adsorb and co-precipitate with calcite precipitates (e.g. Bedore et al., 2008; Golterman, 1995; Olli et al., 2009). Phosphorus retained in the bed sediments can also be released back into discharging waters by a number of processes, which generally include (e.g. Boers and de Bles, 1991; Fox et al., 1986; Søndergaard et al., 1999): 1) the mineralization of organic phosphorus in the bed sediment, 2) increased solubility of phosphate minerals or desorption when external P loads are low and 3) the release of Fe-bound P following the reductive dissolution of Fe^{3+} to Fe^{2+} in anoxic conditions.

Whereas a lot of research has been done on P transport in surface water, little has been done in urban informal settlements especially in SSA. Most research is carried out in agricultural and forested watersheds and in temperate systems (e.g. Blanco et al., 2010; Evans et al., 2004; Rodríguez-Blanco et al., 2013), with very few studies in tropical urban informal systems. However, these two systems could have contrasting mechanisms controlling P transport due to differences in climate, land use and geology. Hence, the fate of P in urban tropical catchments with informal settlements remains unknown (Nyenje et al., 2010). Chua et al. (2009) presented a case study of P transport in a tropical environment but they only focused on P transport dynamics during high and low flows without providing insights into the chemical processes regulating P transport. Informal catchments are rapidly evolving in urban areas in SSA and so is the amount of wastewater and P discharged in the environment. There is therefore a strong need to understand and manage the transport of P in these catchments.

Hence, the main objective of this study was to contribute to the understanding of the processes influencing the transport and fate of P in drainage channels in a slum-dominated tropical catchment in Kampala, Uganda. More specifically, our objectives were to (i) determine the concentrations of the various forms of P discharged from the urban slum catchment during high and low flow conditions, (ii) identify the effect of rainfall runoff on the discharge of P, and (iii) identify the dominant geochemical mechanisms that are likely controlling the fate of P in these channels.

6.2. Catchment description

The upper Lubigi catchment (28 km^2) is located Northwest of Kampala, the capital city of Uganda, with the outlet at latitude 0°21'N and longitude 32°33'E (Fig. 6.1). The catchment is largely urbanized with a number of illegal informal settlements (or slums) such as Bwaise, Mulago, and Kamwokya, located in low-lying areas (Fig. 6.1). Bwaise slum is located at the outlet of the catchment. The underlying geology of the catchment is characterized by Precambrian basement rocks consisting of predominantly granite gneiss overlain by deeply weathered lateritic regolith soils (about 30 m thick) (Taylor and Howard, 1999b). The saturated regolith is an important aquifer containing shallow groundwater flow systems that usually discharge as springs in the valleys of the catchment (Flynn et al., 2012; Taylor and Howard, 1998). These springs generally form the upper reaches or headwaters of the secondary channels or streams (Nyenje et al., 2013b). The mineralogy of the weathered regolith is dominated by non-calcareous kaolinite and quartz minerals with minor amounts of crystalline iron oxide (Flynn et al., 2012). The mean annual rainfall measured at the Makerere University weather station (see location in Fig. 6.1) is 1450 mm/y with two rain seasons (March – May and September – November). During heavy storms, low-lying areas experience a lot of flooding because the catchment is highly urbanized. In slum areas like Bwaise (Fig. 6.1), flooding is even worse because of heavy siltation and blockage of drainage pipes owing to poor solid waste disposal (Kulabako et al., 2010).

The catchment is unsewered (not provided with a sewer). In fact, most parts in the city of Kampala (> 90%) as well as other urban areas in SSA (> 70%) are largely unsewered (Nyenje et al., 2010) implying that most people living in these areas rely on on-site sanitation for wastewater disposal. However, due to poor on-site sanitation systems especially in slum areas (Isunju et al., 2011; Katukiza et al., 2010b; Kulabako et al., 2007; Nyenje et al., 2013b), most wastewater generated in the catchment ends up into the drainage system hence introducing a number of pollutants and nutrients to downstream ecosystems. The drainage system consists of small open drains or tertiary channels located between buildings, which convey a combination of runoff and wastewater (primarily grey water: Katukiza et al., 2010; Katukiza et al., 2014) into a system of larger channels, or secondary channels. The secondary channels then discharge into the primary Nsooba channel (about 3 m wide), which eventually discharges through the Bwaise slum to Lubigi swamp (not shown in Fig. 6.1; see Fig. 3.1). The Lubigi swamp is one of the largest wetlands in the city of Kampala and, like many other wetlands in Uganda, it performs a number of important functions such as the retention of the nutrients derived from urban catchments via drainage channels (e.g. Natumanya et al., 2010; Okiror et al., 2009). Most wetlands in Uganda are, however, being degraded due to extensive encroachment for agricultural activities and infrastructure development. This has hampered their ability to retain nutrients, resulting in deterioration of adjacent water bodies such as Lake Victoria due to eutrophication (Kansiime and Nalubega, 1999b; Kansiime et al., 2005; Kelderman et al., 2007; Kyambadde et al., 2005; Mugisha et al., 2007).

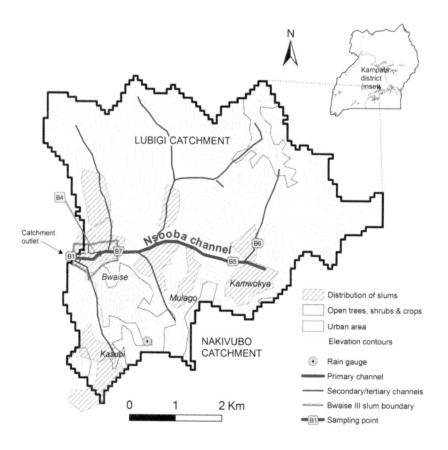

Figure 6.1:   Location of the study area (upper Lubigi catchment) in Kampala, the capital city of Uganda.

6.3. Methodology

6.3.1. Discharge monitoring and precipitation

Stream gauges, each equipped with a Mini-Diver data logger (Schlumberger water services, Delft, the Netherlands), were installed in Nsooba channel at the outlet of the catchment (B1) and at two upstream locations B5 and B6 to monitor discharge along the Nsooba primary channel (Fig. 6.1). The divers continuously recorded water levels at 20 minute intervals. The water levels were compensated for atmospheric pressure using a Baro-Diver (Schlumberger water services, Delft, the Netherlands) installed near the stream gauges. The compensated water levels, H (cm) were converted to discharge, Q (m^3/s) using rating curves ($Q = 0.0006H^2 - 0.0076H : r^2$=0.99, n = 13 for B1, $Q = 0.0039H^{1.4152} : r^2$=0.89, n = 7 for B5 and $Q = 0.0021H^{1.54} : r^2$=0.86, n = 4 for B6), that were developed from a series of discharge measurements carried out. These discharges were measured during different hydrological situations using the salt dilution method (Moore, 2004). It was not possible to perform discharge measurements in the tertiary channel in Bwaise slum (Fig. 6.1). Long-term daily precipitation data, which were used to estimate the annual average precipitation, were provided by the Uganda meteorological department whose nearest weather station is located in Makerere University, about 2 km from the outlet of the catchment (Fig. 6.1). During storms, rainfall data at 5 min intervals and 0.2 mm resolution were obtained from the CREEC project (the Centre for Research in Energy and Energy Conservation; http://creec.or.ug/) located next to the College of Engineering, Design, Art and Technology, Makerere University, about 2 km from the catchment outlet.

6.3.2. Water quality sampling and analysis

We initially collected water samples at intervals of 1 - 2 hrs for 24 hrs during a low flow period on May 26, 2010 (depicting base flow conditions) and during two rainfall events on June 28 and July 28, 2010. The samples were collected in clean 1-litre plastic bottles from the primary channel (Nsooba) at the catchment outlet (B1) and in one of the tertiary channels (B4) in the Bwaise III slum (Fig. 6.1). Tertiary channels are considered to be the primary source of P into the Nsooba primary channel due to sewage effluents especially in slum areas (Katukiza et al., 2010b; Nyenje et al., 2013b). To account for spatial and temporal variability of our data, we collected two more rainfall events on September 18, 2012 and on November 8, 2012 at two upstream locations (B5 and B6) along the Nsooba primary channel (Fig. 6.1).

All water samples were first analyzed on-site for electrical conductivity (EC), temperature, pH, dissolved oxygen (DO) and alkalinity (HCO$_3^-$) immediately after sampling. EC and temperature were measured with an EC electrode (TetraCon 325, WTW) connected to an EC meter (WTW 3310), pH with a pH electrode (SenTix 21, WTW) connected to a pH meter (WTW 3310) and DO with a DO sensor (CellOx 325, WTW) connected to a DO meter (WTW 3310). The meters were calibrated before taking measurements. HCO$_3^-$ was determined by titrating with 0.2M sulphuric acid. After on-site measurements, the water samples were kept in a cool box at 4° C and transported to Makerere University Public Health and Environmental Engineering (PH&EE) Laboratory. Here, the samples were analysed for total phosphorus (TP), orthophosphate (PO$_4$-P) and total dissolved phosphorus (TDP), ammonium (NH$_3$-N), nitrate (NO$_3$-N), total solids (TS) and total suspended solids (SS) in less than 24 h after collecting the samples. TP was determined on unfiltered samples using the ascorbic acid method after digestion with persulfate (APHA/AWWA/WEF, 2005), TDP on filtered samples using the same method as TP and PO$_4$-P determined on filtered samples using the ascorbic acid method (Murphy and Riley, 1962). Particulate phosphorus (PP) was calculated as the difference between concentrations of TP and TDP. Nitrate (NO_3-N) and ammonium (NH_3-N) were determined on filtered samples using the cadmium reduction method and the Nessler method, respectively. All filtered samples were passed through 0.45μm Whatman membrane filters. Final readings were carried out on a HACH DR/4000 U spectrophotometer (USA). Suspended solids (SS) were calculated as the difference between TS and TDS. Total solids (TS) were determined by evaporating an unfiltered sample in an oven at 105 °C for 24 hrs, and then determining the mass of the dry residue per liter of sample, whereas TDS were determined using the same method as TS but on samples filtered using Whatman GF/C filters (APHA/AWWA/WEF, 2005). Samples collected at locations B1 and B4 were very turbid and frequently clogged the filter papers. For these samples, SS were calculated as the difference between TS and TDS estimated from EC as recommended in APHA/AWWA/WEF (2005) in such situations. Here, TDS were estimated from EC using a conversion factor of 0.56, which was computed from a series of TDS and EC values measured during our initial samplings. The factor we used was within acceptable limits (0.55 - 0.7; APHA/AWWA/WEF, 2005). Cations and anions were measured on filtered samples at the UNESCO-IHE laboratory in the Netherlands only for the samples collected at locations B1 and B4: Cations (Ca^{2+}, K^+, Mg^{2+}, Na^+, Mn^{2+} and Fe^{2+}) using an inductively coupled plasma spectrophotometer (ICP - Perkin Elmer Optima 3000) and anions (Cl^- and SO_4^{2-}) by ion chromatography (IC - Dionex ICS-1000). These samples were filtered on-site using 0.45 μm Whatman membrane filters and kept cool at 4 °C until to analysis. Cation samples were preserved by adding 2 drops to concentrated nitric acid.

6.3.3. Mineral saturation indices

We used the PHREEQC model code (version 2, Parkhurst and Appelo, 1999) to calculate the saturation indices of the most important phosphate minerals including calcite ($CaCO_3$), hydroxyapatite ($Ca_5(PO_4)_3(OH)$) and vivianite ($Fe_3(PO_4)_2.8H_2O$) (Stumm and Morgan, 1981). These minerals can regulate P either by precipitation/dissolution (for hydroxyapatite and vivianite) or by co-precipitation (for calcite). Strengite ($FePO_4.2H_20$) was not considered because surface water was anoxic and alkaline (as shown in results), and under these conditions, the strengite forming Fe^{3+} is expected to be insoluble (Appelo and Postma, 2007). Moreover, Fe^{3+} is insoluble in the pH range of 5 – 8. We also determined the saturation indices of rhodochrosite ($MnCO_3$) and siderite ($FeCO_3$) because these metal carbonates can also regulate the concentrations of P by co-precipitation or by precipitation/dissolution reactions.

6.3.4. Sediment sampling and analysis

We collected both surface and deep layer bed sediments at locations B1, B7 and B4 (see Fig. 6.1). The surface layer sediment (herein called shallow sediment) is where most P interactions between the water column and bed sediment occur (Hooda et al., 2000), whereas deep layer sediments represent older deposits that can give insights into earlier interactions that took place. Shallow sediment was loose and was sampled at depths of 0 – 30 cm using a 1m long multi-sampler with a 40 cm internal diameter (Eijkelkamp, the Netherlands). Deep layer sediments were more consolidated and were sampled at depths of 30 – 60 cm using a hand auger. In the tertiary channel (B4, Fig. 6.1), only the shallow sediment was sampled because the channel was lined and the sediment layer was thin (< 10 cm). Due to logistic reasons, suspended sediments were only collected from location B1 during the first rainfall event. This was done by settling and decanting water samples collected in 20 L jerry cans. At the Public Health and Environmental Engineering Laboratory, Makerere University, Uganda, the sediments were air-dried for two weeks. All samples were then sieved using a 2 mm sieve, kept in plastic bags and then transported to UNESCO-IHE, The Netherlands, for analysis. For suspended sediment, only the sample collected during the peak flow at 11:30 AM (event 1) was analysed because the other samples did not contain enough useable sediment for the soil experiments. Sediments were analysed for geo-available metals (Fe, Ca, Mg and K), pH, organic matter (OM) and organic carbon (OC), available phosphorus and grain size distribution. Geo-available metals were extracted with 0.43M HNO_3 (Novozamsky et al., 1993; Rauret, 1998) and analyzed using an inductively coupled plasma spectrophotometer (ICP - Perkin Elmer Optima 3000). Al and Si were not determined due to analytical limitations. Available phosphorus was extracted using the Bray 2 method (Bray and Kurtz, 1945) and analysed by spectrophotometry using the ascorbic acid method. Grain size distribution was determined at the VU University Amsterdam, The Netherlands, by laser diffraction technique using the Helos/KR Sympatec instrument (Konert and Vandenberghe, 1997). The pH was measured on a 2.5:1 water to soil suspension. Soil OC was determined using the Walkley-Black method (Walkley and Black, 1934). Soil OM was estimated from OC using a conversion factor of 1.722 based on the assumption that OM contains 58% carbon (Kerven et al., 2000). All measurements were carried out in duplicate and the results averaged.

6.3.5. Sequential extraction of phosphorus species from selected sediments

To determine the different forms of phosphorus in the stream sediments, we used a sequential extraction technique described by Ruban et al. (2001). The technique was slightly modified to adapt to the equipment available at the laboratory whereby the sediment-solution ratio of 10:1 (mg/ml) was maintained, but the amount of soil used was 500 mg instead of 200 mg. The following forms of P were extracted: P bound to Fe, Al and Mn -oxides and hydroxides (Fe/Al/Mn-bound P), P associated with Ca (Ca-bound P), inorganic P (IP), organic P (OP) and total P. Before analysis, the sediment was first oven dried at 60 °C for 2 hours. For each form of P, extractions then were carried out by adding 50 ml of extracting solution to 500 mg of sediment (or the residue of a previous extraction) and the mixture stirred for 16 hrs. The samples were centrifuged at 4000 rpm for 15 min and P in the extract determined by spectrophotometry using the ascorbic acid method. Fe/Al/Mn-bound P was extracted from 500 mg of dry sediment using 1M NaOH (also referred to as NaOH-P). The residue from this extraction was used for the extraction of Ca-bound P using 1M HCl (also referred to as HCl-P). Total P was extracted from 500 mg of dry sediment using 3.5M HCl. IP was extracted from 500mg of dry sediment using 1M HCl. The residue of the IP extraction was washed with distilled water and calcinated at 450 °C for 3 hrs and then the ash used for extraction of OP using 1M HCl. All extractions were carried out in duplicate and the results averaged.

6.3.6. Phosphorus sorption experiments on selected sediments

Sorption experiments were carried out using duplicate batch experiments. Thereto, 25 mg of sediment samples were accurately weighed and mixed with 500 ml of 0.01 M $CaCl_2$ solution of varying initial P concentrations of 0, 20, 40, 80, 100 and 250 mg/L (i.e. a soil-solution ratio of 0.05). The $CaCl_2$ solution minimizes the competition for sorption sites between phosphate ions and other ions (Froelich, 1988). The phosphorus solutions were prepared using anhydrous KH_2PO_4. The P concentrations used were much higher than those present in the channel ($\cong$ 0.5 mg P/L) in order to establish the maximum adsorption capacity of the sediments. The mixtures were gently shaken on an orbital shaker at 100 rpm to equilibrate. After 24 hours equilibrium time, the final P concentrations in the solutions were measured by spectrophotometry using the ascorbic acid method.

The amount of phosphate sorbed was calculated as:

$$C_{ads} = \frac{(C_o - C_{eqm}) \times V}{m} \times 1000 \qquad (6.1)$$

where C_{ads} is the sorbed amount of P (mg/kg), C_o is the initial P concentration in solution (mg/L), C_{eqm} is the measured P concentration in solution after equilibrium (mg/L), V is the volume of the sample in liters (0.5 L in this experiment) and m is the mass of the dried soil sample (kg).

To establish which sorption isotherm provided the best fit, the Langmuir (Eq. 2) and Freundlich (Eq. 3) equations were fitted to the data. These two equations are often employed to describe adsorption processes (Appelo and Postma, 2007; Golterman, 1995):

$$C_{ads} = \frac{S_{max} C_{eqm}}{K_L + C_{eqm}}, \qquad (6.2)$$

$$C_{ads} = K_F . C_{eqm}, \qquad (6.3)$$

where C_{max} is the maximum adsorbed amount possible (mg/kg) and K_L, K_F and n are adjustable constants (-).

6.4. Results

6.4.1. The hydrochemistry of drainage channels

Table 6.1 presents the hydrochemistry of surface water based on the samples collected at the catchment outlet and in the tertiary channel during base flow conditions. Surface water in the Nsooba primary channel was primarily alkaline (pH = 7.0 – 7.5; mean of 7.3) and had high values of EC (471– 612 µS/cm: mean of 554 µS/cm), high concentrations of HCO_3^- (181 - 326 mg/L: mean of 213 mg/L) and Cl^- (26.2 - 71.3 mg/L: mean of 40.8 mg/L) and relatively high concentrations of cations, primarily Na^+ (4.1 – 51 mg/L; mean of 14.9 mg/L) and Ca^{2+} (3.4 - 25.5 mg/L; mean of 8.2 mg/L). The tertiary channel, which drains the Bwaise slum, was more alkaline (pH = 7.4 – 7.8; mean of 7.7) and had much higher concentrations of dissolved solutes - about 3 times higher than in the Nsooba primary channel. The mean base flow concentration of PO_4-P in the primary channel was 0.36 mg/L (range = 0.11 – 0.78 mg/L) accounting for 30% of TP (1.2 mg/L; Table 6.1). In the tertiary channel, it was 3.3 mg/L (range = 2 – 4.8 mg/L) accounting for 64% of TP (5.2 mg/L; Table 6.1). Hence, PP was the dominant form of P in the primary channel during base flows whereas in the tertiary channel PO_4-P was dominant.

Table 6.1:   Hydrochemistry of drainage channels during base flow conditions at the catchment outlet and in the tertiary drain. Data are shown in concentration ranges with the average values in brackets. Values of variables are given in mg/L except for EC (Electrical conductivity, µS/cm), pH (-) and T (temperature, °C).

Parameter	Nsooba primary channel (Outlet, B1)	Tertiary channel (B4)
EC	471 - 612 (554)	1511 - 1983 (1792)
T	19.7 - 28.4 (24.5)	19.3 - 28.3 (24.2)
pH	7.0 - 7.5 (7.3)	7.4 - 7.8 (7.7)
DO	0.03 - 1.87 (0.7)	0.03 - 0.99 (0.37)
Ca^{2+}	3.4 - 25.5 (8.2)	7.1 - 46.0 (20.3)
Mg^{2+}	0.75 - 6.20 (1.98)	2.3 - 28.3 (7.8)
K^+	2.8 - 24 (7.5)	13.5 - 89.3 (39.5)
Na^+	4.1 - 51.0 (14.9)	21.5 - 229.9 (77.0)
Fe^{2+}	0 - 0.47 (0.12)	0.01 - 0.31 (0.07)
Mn^{2+}	0.14 - 1.28 (0.40)	0.04 - 1.13 (0.48)
NH_3-N	6.8 - 12.1 (10.4)	24.5 - 41.0 (32.4)
NO_3-N	0.5 - 3.6 (1.61)	0 - 7.4 (1.66)
Cl^-	26.2 - 71.3 (40.8)	26.5 - 117.6 (78.3)
HCO_3^-	181 - 326 (213)	522 - 725 (624)
SO_4^{2-}	5.7 - 27.5 (9.5)	3.8 - 28.0 (11.8)
Phosphorus forms		
TP	0.51 - 1.61 (1.15)	3.5 - 7.7 (5.2)
PP	0.09 - 0.93 (0.64)	0.9 - 4.8 (2.2)
PO_4-P	0.11 - 0.78 (0.36)	2 - 4.8 (3.3)

The channels were generally anoxic and characterized by low concentrations of dissolved oxygen (DO < 1 mg/L; Table 6.1) in both the primary and the tertiary channel. Consequently, the concentrations of NO_3-N were low (< 1.6 mg/L) while NH_3-N concentrations were high (mean of 10.4 mg/L in the primary channel and 32.4 mg/L in the tertiary channel). With regard to redox sensitive elements, Fe^{2+} concentrations were very low (mean < 0.12 mg/L) whereas SO_4^{2-} concentrations were high (about 10 mg/L; Table 6.1). Although Mn^{2+} concentrations were also low (mean of 0.4 mg/L in the primary channel and 0.48 mg/L in the tertiary channel; Table 1), they were relatively higher than the Fe concentrations. Hence, the redox state of the surface water was likely Mn-reducing.

6.4.2. Saturation indices (SIs)

Surface water is considered to be saturated or near saturated when the saturation index (SI) of a given mineral ranges between -0.5 < SI < 0.5 (Deutsch, 1997). A range of -0.7 < SI < 0.7 is also considered in some studies (e.g. Griffioen, 2006). In the tertiary drain, most samples were saturated (-0.4 < SI < 0.4) with respect to calcite and rhodochrosite (Fig. 6.2b). In the primary channel, some samples were near saturation levels with respect to calcite and rhodochrosite (0 < SI < -0.7) whereas some were undersaturated (SI < -1) especially during rain events (Fig. 6.2a).

The two minerals (calcite and rhodochrosite) therefore likely regulated P concentrations by co-precipitation (or the adsorption of P onto carbonate precipitates). The SI values of $MnHPO_4$ were high but relatively constant (SI ranged from 2 to 3) in both the tertiary and the primary channels and in all the events (Fig. 6.2). These high saturation levels of $MnHPO_4$ were likely regulated by the near saturated levels of rhodochrosite, which suggests that $MnHPO_4$ was not reactive. The term reactive mineral here refers to a mineral that can easily dissolve into or precipitate from the water column under certain conditions (Deutsch, 1997). The SI values of Fe phosphates (vivianite) were most of the times less than -1 (undersaturated) implying that these minerals were either not present or not reactive. Surface water was supersaturated with respect to hydroxyapatite in the tertiary channel (Fig. 6.2b), but undersaturated in the primary channel (Fig. 6.2a).

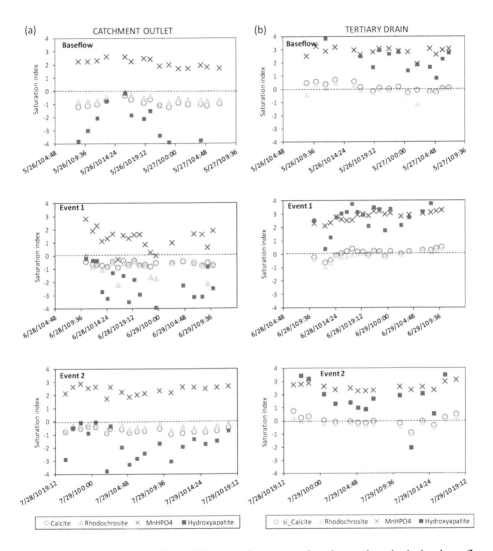

Figure 6.2: Saturation indices of the most important phosphate minerals during base flow and runoff events: (a) at the outlet of the catchment and (b) in the tertiary channel draining the Bwaise III slum. The minerals vivianite and siderite are not shown here because they had high negative values of saturation indices suggesting that they were not present.

6.4.3. Phosphorus concentrations during base flows

During base flow conditions, high concentrations of P were measured in the Nsooba channel, with values ranging from 0.51 to 1.61 mg/L (average 1.15 mg/L) for TP, 0.09 to 0.93 mg/L (average 0.64 mg/L) for PP and 0.11 to 0.78 mg/L (average 0.36 mg/L) for PO_4-P (Table 6.1, Fig. 6.3a). The dominant form of P was PP accounting for 56% of TP whereas PO_4-P accounted for 31%. Concentrations of P seemed to vary slightly during the day with relatively higher concentrations (> 1 mg/L as TP) occurring between 7: 00 AM - midnight, and lower concentrations occurring after midnight (< 1 mg/L TP) (Fig. 6.3a). The base flow P concentrations in the tertiary channel were much higher than in the primary Nsooba channel and ranged 3.5 – 7.7 mg/L (average of 5.2 mg/L) for TP, 0.9 – 4.8 mg/L (average of 2.2 mg/L) for PP and 2 - 4.8 mg/L (average of 3.3 mg/L) for PO_4-P (Table 6.1; Fig. 6.6a). Particulate P accounted for 42 % of TP whereas PO_4-P accounted for 58%.

6.4.4. Phosphorus concentrations during rainfall events

We collected hourly data over 24 hours on four (4) rainfall events on 26 June 2010 and 28 July 2010 (at the catchment outlet, B1, and in the tertiary channel, B4) and on 18 September 2012 and 8 November 2012 (at upstream locations B5 and B6 along the primary channel). In all rainfall events, there was a simultaneous increase in concentrations of TP and PP (and SS as well) with peak concentrations almost coinciding with the peak discharge of the rainfall-runoff hydrographs (Figs. 6.3 - 6.5). Thereafter, base flow concentrations were restored.

At the catchment outlet, B1, about 8.6 mm (intensity of 14 mm/h) fell during the first event (28-29 June 2010) producing a peak discharge of 6.7 m^3/s. Subsequently, there was an increase in concentrations of TP and PP and concentration peaks of 4 mg/L for TP and 3.66 mg/L for PP (about 92% of TP) were realized (Fig. 6.3b). During the second event (28 – 29 July 2010; 14.8 mm, intensity of 6.5 mm/h), two smaller peak discharges of 1.3 and 1.4 m^3/s were produced. Consequently, two peak concentrations of TP and PP were produced (Fig. 6.3c). The first concentration peak had 3.0 mg/L for TP and 2.4 mg/L for PP whereas the second had 2.1 mg/L for TP and 1.5 mg/L for PP.

In the upper locations (B5 and B6; Fig.1), the drainage area was smaller (about 8 km^2) resulting into smaller discharges (peak of about 2 m^3/s; see Fig. 6.4 and Fig. 6.5) after a storm event. However, the peak concentrations of PP and TP at these locations after the storm events were slightly higher than those observed at the catchment outlet, which had a larger drainage area. For example, at location B5, about 7 mm of rain fell on 18 September 2012 (event 3) producing a peak discharge of 1.9 m^3/s. The resulting peak concentrations of TP and PP were 6.8 mg/L for TP and 5.7 mg/L for PP (about 84% of TP) (Fig.4a).

In the tertiary channel, the P trends following storm events were similar to those in the primary channel, except here, the concentrations were much higher (see Fig. 6.6b and Fig. 6.6c). For example, after the first rainfall event (28 June 2010), peak concentrations of 19.7 mg/L for TP and 14.1 mg/L for PP were realized (almost 3 times higher than corresponding peak concentrations in the primary channel) (Fig. 6.6b).

Rainfall events also increased the concentrations of suspended solids (SS) and we observed that the concentration peaks of SS generally coincided with those of PP and TP (see Fig. 6.3 to Fig. 6.6). However, the responses of PO_4-P were not readily evident as was observed for TP and PP (Figs. 6.3 – 6.5) except in the tertiary channel (Fig. 6.6b and Fig. 6.6c). In all runoff events sampled, the concentrations peaks of PP and TP were realized after the peak discharge implying a late delivery of nutrients.

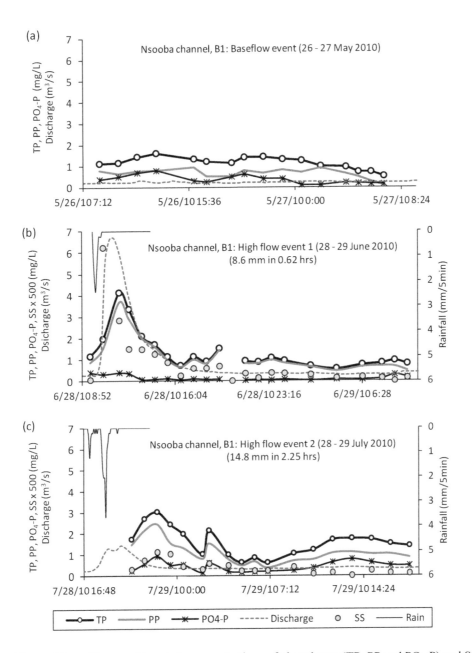

Figure 6.3:   Temporal trends in concentrations of phosphorus (TP, PP and PO$_4$-P) and SS in Nsooba channel at location B1 (catchment outlet) during (a) a base flow event, (b) rainfall event 1 and, (c) rain fall event 2.

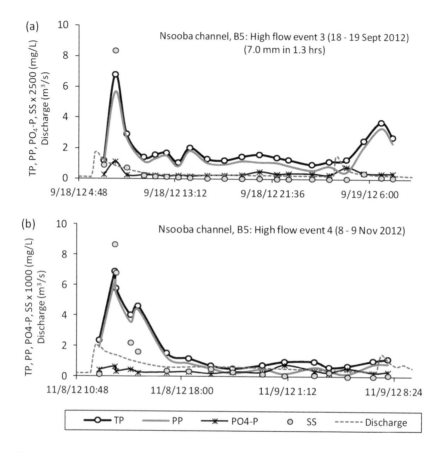

Figure 6.4:   Temporal trends in concentrations of phosphorus (TP, PP and PO$_4$-P) and SS in Nsooba channel at location B5 during (a) rainfall event 3 and (b) rainfall event 4. Note: During these events, precipitation data was not available and is therefore not presented.

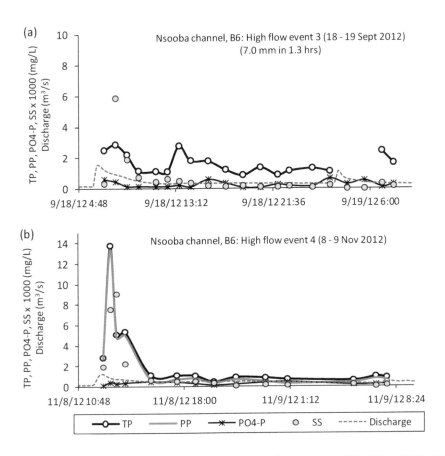

Figure 6.5:   Temporal trends in concentrations of phosphorus (TP, PP and PO$_4$-P) and SS in Nsooba channel at location B6 during (a) rainfall event 3 and (b) rainfall event 4.

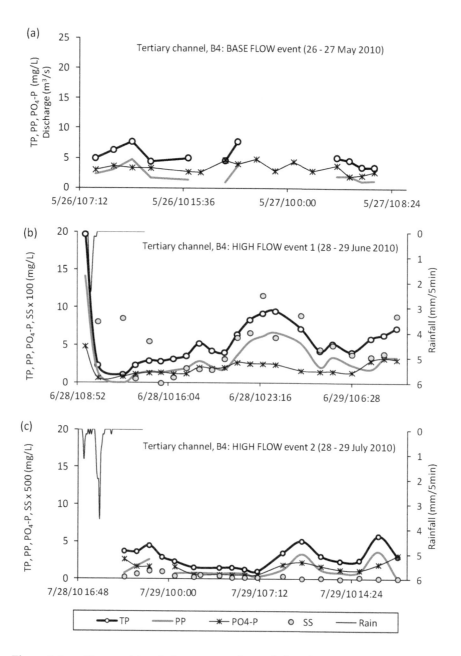

Figure 6.6: Temporal trends in concentrations of phosphorus (TP, PP and PO₄-P) and SS in a tertiary channel discharging Bwaise slum during (a) a base flow event, (b) rain event 1 and, c) rain event 2. Discharge measurements were not possible at this site.

6.4.5. Physical and chemical characteristics of sediments

Table 6.2 presents the results of the soil analyses. The bed sediments had very high values for sand content (63 – 83%) and low values for silt content (5 – 11%). Suspended sediments, however, had a high silt content (56%) and low sand and clay contents (< 23% each). The sediments were alkaline with pH ranging from 7.1 to 7.3. Organic matter content ranged from 1.8 to 3% and was highest in deeper sediments. Based on the 0.43 M HNO_3 extraction, Ca was the dominant cation followed by Fe and then Mn. Al was not determined due to analytical limitations. Hence Ca, Fe and Mn were most capable of interacting with P. Suspended sediments contained higher contents of Ca, Fe and Mn than the bed sediments.

The results of sequential P extraction showed that the total sediment P ranged from 1375 to 1850 mg/kg. Inorganic P was the dominant form of P in the sediments accounting for over 63% (range of 64 – 80%) of the total sediment P (Fig. 6.7). Here Ca-bound P and Fe/Mn-bound P were present in almost equal proportions (i.e. 51 - 54% and 46 – 49% of total P, respectively). The measured IP contents (not shown) were close to the calculated values (sum of Ca-P and Fe-P) except for suspended sediments. Organic P accounted for 17 – 22% of total sediment P whereas adsorbed P (i.e. Bray-2 extractable P) accounted for only 5 – 8% of total sediment P. Hence, P retained in sediments was probably a result of precipitatation of Ca and Fe phosphates or the adsorption of P to Fe/Mn-oxides. Suspended sediment collected during the peak of the first rain event had the highest content of phosphorus (total sediment P of 2316 mg/kg) of which Ca-bound P accounted for 43% of the total sediment P and Fe/Mn-bound P accounted for 29% of the total sediment P. There seemed to be no significant differences between the shallow and deep sediments.

From the P sorption experiments, the Langmuir isotherm provided the best fit of the sediment data (r^2 = 0.85 - 0.98; Fig. 6.8). The sediments in the tertiary channel, however, had a poor fit (r^2 = 4.3) probably due to the relatively low Fe content (2262 mg/kg) compared to that in the Nsooba channel (> 3400 mg/kg) (Table 6.2). The maximum sorption capacity (C_{max}) of the sediments ranged from 820 to 2350 mg/kg (Fig. 6.8). Deeper sediments had the highest sorption capacity (C_{max} = 2350 mg/kg). Shallow sediments generally had low sorption capacities (C_{max} = 1550 mg/kg in Nsooba sediment and C_{max} = 850 mg/kg in the tertiary channel). From Fig. 6.8, it becomes clear that the predicted amount of P sorbed to the sediment based on the measured PO_4-P concentrations (means of 0.36 and 3.3 mg/L; Table 6.1) was about 0.02 mg/g (or 20 mg/kg), which is much less than the sum of the inorganic and the Bray-2 extractable P (1.15 – 1.29 mg/g; Fig. 6.7). Hence, the sediments were likely P saturated.

Table 6.2: Physical and chemical properties of bed and suspended sediments in the Nsooba channel and in the tertiary channel of Bwaise slum.

Location	pH (-)	OM* (%)	OC* (%)	Geo-available metals (mg/kg)					Grain size distribution** (%)		
				Ca	Mg	K	Mn	Fe	Sand	Silt	Clay
Deep sediment (30 - 60 cm)											
B1 (outlet)	7.1	3.0	1.7	3636	173	105	387	3761	79	14	7
B7	7.3	3.0	1.7	4691	216	194	348	3269	77	16	7
Shallow sediment (< 30 cm)											
B1 (outlet)	7.3	2.0	1.2	8505	375	311	767	3437	63	26	11
B7	7.2	2.7	1.6	2729	130	117	375	2338	84	11	5
B4 (tertiary drain)	7.3	1.8	1.1	7620	292	381	379	2262	83	12	5
Suspended sediment (Event 1 at 11: 30 am)											
B1 (outlet)	-	-	-	18125	939	1405	908	5755	23	56	21

*OM, organic matter; OC, organic carbon; ** Clay (< 2μm), silt (2 μm - 50 μm) and sand (50 μm - 2mm)

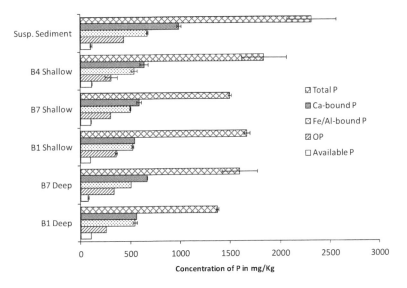

Figure 6.7:   Mean concentrations of P forms in the bed and suspended sediments. The P forms included Fe/Al-bound P, Ca-bound P, OP (Organic P), Total P and available P (Bray-2 extractable P). Suspended sediments were only collected at peak flow during Event 1 (Error bars represent standard error, n = 2).

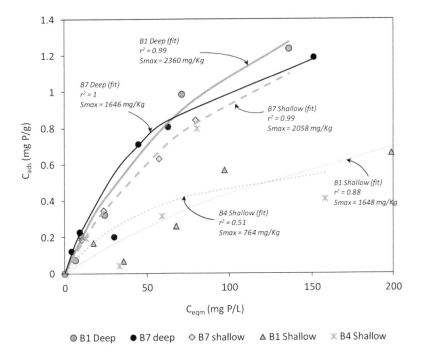

Figure 6.8:   Langmuir sorption isotherms for sediments sampled at locations B1 (catchment outlet), B7 (inlet of slum area) and B4 (tertiary channel).

6.5. Discussion

6.5.1. Phosphorus exported from the catchment

Our results showed that the average base flow concentrations of PO_4-P and TP at the outlet of the studied catchment were 0.36 and 1.2 mg/L, respectively (Fig. 6.3a; Table 6.1). These concentrations are about 16 times the eutrophication limit of 0.075 mg TP/L proposed by Dodds et al. (1998) for streams implying that the channel was very eutrophic. For a mean base flow of 0.22 m^3/s (Fig. 6.3), the fluxes of PO_4-P and TP from the studied catchment were about 0.3 and 0.95 kg km^{-2} d^{-1}, respectively. These fluxes appear to be very high compared to those normally reported in published literature for agricultural, forested and other urban catchments. Jordan et al. (2007) for example reported a TP flux of 0.2 kg km^{-2} d^{-1} for a 5 km^2 rural agricultural catchment in Northern Ireland. In a mixed land use catchment in Galicia, Spain, Rodríguez-Blanco et al. (2013) reported a much lower TP flux of 0.04 kg km^{-2} d^{-1}. Zhang et al. (2007) found a TP flux of only 0.034 kg km^{-2} d^{-1} in nutrient runoff from forested watershed in central Japan. Nhapi et al. (2006) showed that the TP flux for the two major river inflows to the eutrophic Lake Chievero in the city of Harare (Zimbabwe) ranged between 0.1 and 0.35 kg km^{-2} d^{-1}. Although these rivers were heavily polluted by on-site sanitation and sewage overflows from treatment plants, the TP fluxes were still low when compared to those in our study. Our findings therefore suggest that phosphorus exported from urban catchments with informal settlements (or slums) poses a very serious threat to downstream surface water quality due to eutrophication.

6.5.2. Source of phosphorus

There are three possible sources of the high concentrations of P observed in the Nsooba channel: groundwater exfiltration, precipitation and anthropogenic sources (use of agricultural fertilizers and sewage effluents). However, groundwater and precipitation in the study area contain relatively low concentrations of P (< 0.06 mg/L as PO_4-P; Nyenje et al., 2013b). There is also limited use of fertilizers for agriculture in the study area and in most parts of Uganda. Hence, sewage effluents were the most likely sources of P in the studied channels. The lack of a sewer system and the existing poor on-site sanitation systems in the studied catchment means that most wastewater generated from households was directly discharged untreated into the drainage channels thereby introducing high levels of nutrients. The water quality of these channels was characterized by high concentrations of EC, HCO_3^-, NH_4^+, PO_4-P and cations (Table 6.1), which is typical of wastewater streams derived from sewage effluents. A recent study by Katukiza et al. (2014) revealed that most wastewater in these channels is composed of grey water (wastewater from bathrooms, kitchens and laundry), which is normally transported to the primary channel via small channels (or tertiary channels) between households especially in slum areas. We monitored one of the tertiary channels in the Bwaise slum and indeed it contained very high P concentrations (TP = 3.5 - 7.7 mg/L and PO_4-P = 2 - 4.8 mg/L; Table 6.1) similar to those found in grey water in this slum area (6 - 8 mg TP/L; Katukiza et al., 2014). The influence of domestic wastewater as a source of P can also be seen from the 24 hr trends of the base flow concentrations of TP in the primary channel (Fig. 6.3a). From midnight onwards, the concentrations of TP in the primary channel were less than 1 mg/L, whereas from 7:00 am to 10:00 pm, they were greater than 1 mg TP/L (Fig. 6.3a). This suggested that P was being added to the system during the hours of the day and this was likely due to wastewater effluents from human activity. This may also explain why ortho-phosphates accounted for the largest proportion of P in the tertiary channel given that ortho-phosphate is the dominant form of P in wastewater stream

(Bedore et al., 2008). Therefore, we concluded that sewage effluents, especially from grey water streams, were the main source of P in the studied catchment.

6.5.3. Effects of rainfall runoff events

Rainfall-runoff events indicated a flushing of P predominantly in particulate form (e.g. 92% of TP during event 1 was particulate, Fig. 6.3b). This flush was characterised by a sharp increase in concentrations of TP and PP (and SS) during the rising limb of the hydrograph and a subsequent decrease during the falling limb and finally restoration of base flow concentrations. This phenomenon was present in all the four rainfall events at the catchment outlet (B1), the upstream stations B5 and B6 and in the tertiary channel (B4). During the second event (Fig. 6.3c), there were two discharge peaks at B1 which resulted in two concentration peaks of TP and PP, with the later peak lower than the former probably due to flushing effects. The peak concentrations of SS during the events coincided with those of TP and PP in all the four rain events (Fig. 6.3 to Fig 6.6), which implied that most phosphorus transported during rain events was associated with suspended sediments. These sediments can be mobilised from the resuspension of the P-rich bed sediments accumulated in the channel or by erosion of material stored on the urban surface. We observed that the peak concentrations of TP and PP in the primary channel during storm events were not so different at the catchment outlet and at the upstream locations B5 and B6, yet discharge varied significantly because of differences in drainage areas. It is therefore unlikely that catchment runoff was the primary source of TP and PP during rain events. Instead, we think that resuspension of P-rich bed sediment played a more important role. This sediment phosphorus was likely deposited during low flows because a large amount of P (about 56% of TP) in the primary channel was particulate during the baseflow event sampled. Our results therefore suggest that there was a flushing of TP and PP, primarily due to the resuspension of bed sediments. Flush effects were not observed for the dissolved form of P (PO_4-P) probably because other mechanisms such as precipitation/dissolution or dilution were more dominant.

First-flush effects for PP and SS have been reported in many studies investigating P transport during storm events (e.g. Stutter et al., 2008; Zhang et al., 2007). They occur when the rising limb of a hydrograph contains higher concentrations of pollutant than the falling limb (Deletic, 1998). Results from our study, however, seem to suggest that PP and TP did not exhibit first-flush effects. This is because the concentration peaks of PP and SS were most of the times realised after the peak events (see Fig. 6.3 to Fig. 6.6), implying that the falling limb contained higher concentrations of pollutants than the rising limb. This could have been caused by the poor on-site sanitation systems in the catchment. Here, wastewater especially from pit latrines is normally released into drainage channels after rain events (particularly when increased flows are observed) as a cheap way of emptying the latrines (see description study area). One study by Chua et al. (2009) also investigated a tropical catchment with proportions of rural and urban land use similar to the catchment we studied, and they also observed that the first-flush effects were generally weak for TP and PP. In our study, however, our sample collection was not frequent enough during peak flows and it is therefore not possible to confirm whether the first-flush effects for PP were present or not.

6.5.4. Sediment-water column phosphorus interactions

P retention and role of overlying water

The settling of particulate matter is one of the most important mechanisms of P retention in the channel bed sediments (Reddy et al., 1999). As already mentioned above, our results showed that most P discharged at the outlet of the catchment during low flows was particulate (56% of TP) and was likely retained along the channel bed by settling. Fractionation analyses also showed that the P retained in the bed sediment was largely inorganic accounting for 64 - 80 % of the total sediment P. Additionally, inorganic sediment P was mostly bound to Ca (HCl-P) and Fe/Mn (NaOH-P) and in almost equal proportions (51 - 54% and 46 - 49 % respectively; Fig. 6.7). This implies that P retention processes in the bed sediment were attributed to either mineral precipitation or the adsorption of P to sediments. Therefore, sediment inorganic P that was bound to Ca (or Ca-bound P) could either be mineral precipitates of hydroxyapatite or P adsorbed to calcite. These two minerals, hydroxyapatite and calcite, are widely reported in literature to regulate P transport in river systems by precipitation and co-precipitation respectively (e.g. Golterman, 1995; Griffioen, 2006; Reddy et al., 1999; Tournoud et al., 2005). However, the nature of the rocks in our study area (Precambrian granite-gneiss rocks; see section 6.2) shows that there are no carbonate-bearing rocks like calcite. Hence, the likely source of calcite in sediment, if it was present, was through mineral precipitation reactions in the water column (i.e. calcite precipitates). Sediment inorganic P bound to Fe/Mn (or Fe/Mn-bound P) was likely a result of the adsorption of P to Fe-oxides or the precipitation of Fe and Mn phosphates. Note that we were unable to determine Al in this study and therefore the role of Al-oxides to the adsorption of P is not considered. Here, below, we try to identify the relative importance of these two processes (sorption and mineral solubility) to the retention of P in the sediment.

With regard to Ca-bound P, geochemical speciation results showed that surface water was most of the times near saturation or saturated with respect to calcite in both the primary and the tertiary channels ($0 < SI < 0.7$; Fig. 6.2) implying that calcite was reactive (may equilibrate with surface water) in the studied channels. The channels were, however, undersaturated with respect to hydroxyapatite (SI generally less than -1) in the primary channel but oversaturated in the tertiary channel (SI generally greater than 1). This phenomena likely occurred due to the high concentrations of Ca^{2+} (7.1 – 46 mg/L) and phosphate (2 – 4.8 mg/L as PO_4-P) in the tertiary channel (Table 6.1) and the low concentrations in the primary channel (Ca^{2+} = 3.4 – 25.5 mg/L; PO_4-P = 0.1 – 0.9 mg/L) following dilution. Hence, hydroxyapatite seemed not to be reactive. Indeed some studies suggest that hydroxyapatite precipitation can only take place when the SI > 9.4 and when Ca^{2+} concentrations are very high (> 100 mg/L) (e.g. Diaz et al., 1994). Given that phosphorus has strong adsorption affinity to calcite (e.g. Reddy et al., 1999) and that calcite was reactive, we conclude that Ca-bound P in the bed sediments was due to the adsorption of P to calcite precipitates.

With regard to Fe/Mn-bound P, the studied drains were undersaturated with respect to vivianite (SI < -3; not shown in Fig. 6.2) implying that vivianite was not present in the studied channels. Strengite was also likely not present because Fe^{3+} is insoluble in the pH range of 5 – 8 (Appelo and Postma, 2007), which was the pH range of the drainage channels studied (pH = 7.0 – 7.8). $MnHPO_4$ was consistently oversaturated (SI $\cong$ 2 – 3) in both the tertiary and the primary channel, which suggests that this mineral was not reactive. It is also possible that the high SI values of $MnHPO_4$ were caused by the near saturated state of rhodochrosite ($MnCO_3$) because surface water was saturated with respect to this mineral (Fig. 6.2a and Fig. 6.2b). Although mineral carbonates such as calcite can to regulate P by co-precipitation (e.g. Freeman and Rowell, 1981), there are currently no published reports of phosphorus scavenging by rhodochrosite ($MnCO_3$). Hence, P bound to the Fe-/Mn-oxides was likely due to the adsorption of P to Fe-/Mn-oxides. We think that Fe-oxides played a more important role instead of Mn-oxides because of the dominance of Fe-rich laterite in the study area. The adsorption of P to Fe-oxides likely occurred during and after rain events when there was resuspension of bed sediment and the erosion of Fe-rich soils from urban surface.

Nutrient ratios are also often used to predict whether the deposited sediment P is produced by adsorption or precipitation reactions or both. A molar ratio of Fe/P $\cong$ 2 (range = 1.5 – 2.5) suggests that Fe-bound P in the sediment was produced by the precipitation of Fe-phosphates such as vivianite and strengite whereas higher ratios (Fe/P = 3.3 – 9.7) suggest that it was produced by the adsorption of P to Fe-oxides (e.g. Clark et al., 1997; Cooke et al., 1992; Gunnars et al., 2002). For Ca, the limiting Ca:P ratio for mineral precipitation is about 1.7 (Freeman and Rowell, 1981). In our study, both the Fe:P and Ca:P molar ratios in the sediment were too high (> 3.6 for Fe:P and > 5.2 for Ca:P; Table 6.3) to argue that the precipitation of Fe and Ca phosphates took place. Instead, these high ratios confirm that sediment inorganic P was produced by the adsorption of P to calcite and Fe oxides. These revelations are in agreement with our earlier arguments from mineral saturation indices that the adsorption of P to calcite and Fe-oxides regulated P transport in the studied drains.

Table 6.3:   Nutrient and metal molar ratios of bed sediments and the water column in comparison with literature values.

System	Fe:P* (molar)	Ca:P** (molar)	OC:OP (molar)	Remarks
Nsooba channel, Uganda				
Sediment B1 - deep	3.8	5.0	17	This study
Sediment B7 - deep	3.6	5.2	15	This study
Sediment B1 - shallow	3.6	12.1	9	This study
Sediment B7 - shallow	2.6	3.6	14	This study
Water column (baseflow)	0.2	17.8	-	This study
Suspended sediment	14.9	4.5	-	This study
Tertiary channel, Uganda				
Sediment B4 - shallow	2.3	9.2	9	This study
Water column (base flow)	0.0	4.7		This study
Nakivubo channel, Uganda				
Bed sediment	1 – 6		68 – 92	Kansiime & Nalubega (1999)
Mozhaisk reservoir, Russia				
Bed sediment	5.7			Martynova (2011)

P*, Iron bound P (NaOH - P); P**, Calcium bound P (HCl - P)

Based on the discussions above, Fig. 6.9 shows a schematic of the processes likely affecting P transport in the studied catchment.

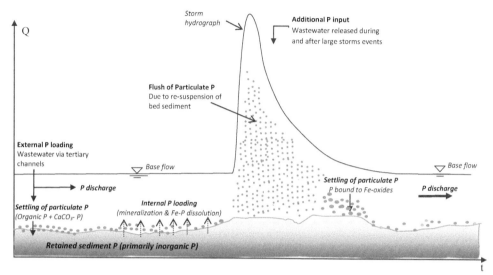

Figure 6.9:   Schematic of the possible phosphorus transport processes during low flow and high flood events. The grey oval shapes of different sizes illustrate settling and resuspension of particulate P.

Currently, there are few published studies in urban informal settlements in SSA with which we can compare our results with regard to P transport. One study by Kansiime and Nalubega (1999), however, did try to investigate the removal of P by sediment in the Nakivubo channel/swamp, which had received wastewater from the city of Kampala (Uganda) for over 30 years. The findings of this study, on the contrary, suggested that the precipitation of $CaCO_3$ was not a very important process for P retention because of the low Ca content (60 mg/kg) and the low Ca-bound P (10% of TP) in the sediment. The retention of P was instead attributed to precipitation of vivianite because of the low Fe:P molar ratios (1 - 3). However, this study did not carry out geochemical speciation of phosphate phases in the overlying water to confirm if there was a strong likelihood for vivianite to precipitate. In our study, Fe:P ratios in the sediment were generally high (> 3.6; Table 6.3) and the overlying water was undersaturated with respect to Fe phosphates. In addition, the sediment Ca content was very high (> 2000 mg/kg; Table 6.1) while at the same time Ca-bound P contributed over 50% of total sediment P. This implies that there were indeed interactions between Ca and P in our study. These findings show that P retention processes in surface water in urban informal settlements can vary significantly depending on the location

P release from sediments

Our results further indicated that the bed sediments, particularly the shallow ones, were P saturated because the total sediment P was almost equal to or more than maximum P sorption capacity, C_{max} (Fig. 6.7; Fig. 6.8). For example, the C_{max} of the shallow sediments at location B1 was 1550 mg/kg whereas the total sediment P was 1668 mg/kg (Fig. 6.7; Fig. 6.8). In the tertiary drain, the C_{max} was 820 mg/kg whereas total sediment P was 1840 mg/kg. The P saturation state of sediments is one of the most important factors that indicate the potential for sediments to release P (Hooda et al., 2000). Three processes are mainly responsible for the release of P from sediments: 1) mineralization of organic matter, 2) desorption, and 3) reductive dissolution of P bound to Fe-oxides (e.g. Boers and de Bles, 1991; Fox et al., 1986; Søndergaard et al., 1999). Given that surface water was Mn-reducing, it is unlikely that reductive dissolution of Fe-bound P took place. Hence, if P release took place, it was by mineralization of organic matter or by desorption.

Metal/nutrient and C/P molar ratios are also often used to indicate P release by dissolution of metal bound P and mineralization of organic P respectively. Fe:P (and Ca:P) ratios < 2 indicate a tendency for bed sediments to release PO_4^{3-} (Jensen et al., 1992). Table 6.3 shows that Fe:P ratios in the water column in our study were very low (< 2) whereas Ca:P ratios were very high (up to 17) indicating that P was potentially released from Fe-bound P in the sediment by desorption. On the other hand, P release by mineralization usually occurs when the C:P molar ratio < 200 (Stevenson, 1986). In our study, the C:P ratios of all sediments collected were less than 200 implying that there was a potential for mineralization of P from the organic P retained in the bed sediment.

This study has provided useful insights into the processes regulating the transport of sanitation-related phosphorus in drainage channels in a typical urban slum catchment. We showed that the adsorption of P to Fe-oxides in the sediments and to calcite precipitates in the water column played an important role in regulating P transport to downstream areas. We have also demonstrated the presence of flush effects and the role of the channel-bed sediments to release phosphorus back into the water column. Knowledge of these processes is crucial in developing process-based water quality models, which can aid policy and decision making regarding strategies to reduce nutrients exported from urban catchments. A complete understanding of these processes, however, requires more research to be carried out in these types of catchments. Future work could focus on using on-site automated samplers and analysers in order to obtain high-resolution data, to enable improved understanding (at a much higher scale) of the P transport processes and the trends that occur when hydrological conditions change.

6.6. Conclusions

In this study, we attempted to understand the processes governing the transport and retention of phosphorus (P) during high and low flows in a channel draining a 28 km^2 urban catchment with informal settlements in Kampala, Uganda. Our results revealed the following:

- A large amount of phosphorus was discharged from the studied catchment. The base flow concentrations of P in the primary channel were 1.15 mg/L for total P and 0.36 mg/L for PO_4-P, which were about 16 times the minimum required to cause eutrophication. The corresponding P fluxes of 0.3 kg km^{-2} d^{-1} for PO_4-P and 0.95 kg km^{-2} d^{-1} for TP were also of several orders of magnitude higher than values normally reported in published literature for other catchments.

- By comparing the hydrochemistry and P concentrations in the primary channel and a tertiary channel draining a slum area, we were able conclude that the primary source of P in the channels was the direct discharge of untreated wastewater into the tertiary channels, primarily due to grey water effluents from informal settlements.

- In the four (4) rain events we sampled, we observed a flushing of P mainly in particulate form. We attributed this flushing to the resuspension of the P-rich bed sediments that had accumulated in the channel during low flows. It is unlikely that the terrestrial runoff significantly contributed to the flushing of TP and PP because the concentration peaks of TP and PP during the rain events were almost similar irrespective of the sample location along the primary drain. Our results, however, seemed to suggest that first-flush effects were not present. In all the rain events sampled, the concentration peaks of TP, PP and SS were realized after the peak discharge. This was likely caused by the poor on-site sanitation practices in the catchment whereby most residents, especially in informal settlements, release untreated wastewater into drainage channels during and after storm events.

- In relation to chemical processes, our results indicated that P transport in the channels was regulated by the co-precipitation of P with calcite precipitates and by the adsorption of P to Fe-oxides, especially during rain events when there was resuspension of sediments. These findings were consistent with our other findings that that the retained P in the sediment was largely inorganic (64 – 80 % of total sediment P) and was bound to Ca and Fe/Mn-oxides in almost equal proportions (51 – 54% and 46 – 49 % respectively). The retention of P by settling of organic matter also seemed to be important because organic-bound P in the bed sediments was also relatively high (i.e. 17 – 22% of the total sediment P). The sediments, however, showed potential to release P to the overlying water by mineralization of organic matter and desorption of P bound to Fe-oxides/hydroxides.

Our study provides useful insights into mechanisms likely controlling P transport in a typical urban catchment with informal settlements. To have a complete understanding of the P transport processes in these catchments, we recommend that additional P transport studies be carried out in other urban informal catchments with emphasis on high-resolution nutrient monitoring during high flows.

Chapter 7

Conclusions, recommendations and future research needs

7.1. Conclusions and recommendations

Relevance of the study and research approach

In most large cities in sub-Saharan Africa (SSA), urbanization is synonymous with development of urban slums (or informal settlements), primarily due to poor planning and lack of proper management systems. These informal settlements are often unsewered and lack access to proper on-site sanitation systems (collection and treatment of waste at the source of generation), hence resulting in the indiscriminate disposal of untreated or partially treated wastewater to groundwater and surface water systems. Wastewater contains a lot of nutrients and therefore the uncontrolled disposal of wastewater to the environment can result in excessive discharge of nutrients to downstream water resources leading to deterioration of surface water resources due to eutrophication. Eutrophication is the increased primary production of a lake system caused by enrichment of nutrients (principally nitrogen, N and phosphorus, P) (Smith et al., 1999). It is characterized by excessive growth of the water hyacinth, blooms of green algae including the toxic blue-green algae, depletion of oxygen levels, bad smells, and subsequent death of aquatic organisms such as fish. Many fresh water lakes in cities in SSA are faced with problems of eutrophication due to the excessive discharge of nutrients from poorly-sanitized catchments to the environment (chapter 1 and 2). However, the mechanisms by which these nutrients are transported from the sources of generation to downstream water resources via groundwater and surface water flow paths have not been adequately researched in urban areas in SSA (see chapter 1 and 2).

Hence, the aim of this thesis was to investigate the processes governing the transport of sanitation-related nutrients (N and P) in groundwater and surface water in an unsewered urban slum area. The study area was Bwaise III parish slum (low-lying and largely reclaimed from a wetland) and its catchment (the Lubigi catchment) in Kampala, Uganda. This study is relevant because it addresses the need to improve the water quality status of urban fresh water resources in SSA, eventually improving drinking water security. The research was mainly guided by four specific objectives: (1) to carry out a critical review of the literature on eutrophication and nutrient pollution in SSA, (2) to identify the sources of nutrients in these areas 3) to estimate the nutrient inputs from on-site sanitation systems to shallow groundwater, and 4) to identify the dominant processes governing the transport and fate of these nutrients in groundwater and surface water. The study applied an experimental, a modeling and a processes-description approach, with the later the major focus of the study. This multi-methods approach involved carrying out extensive field and laboratory investigations along the major pathways of nutrients to downstream ecosystems (i.e. groundwater, surface water and precipitation) in order to obtain hydrochemical data to describe the nutrient transport processes. The study was carried out at both the catchment scale and at micro-plot scale to be able to understand the entire cycle of nutrient transport processes in the urban hydrological cycle. The focus was on the essential nutrients that cause eutrophication and these were nitrogen (N) in the form of ammonium (NH_4^+) and nitrate (NO_3^-), and phosphorus (P) in the form of ortho-phosphate (PO_4^{3-}). In the following text, the findings of the study are summarized:

Eutrophication in urban areas in SSA: research needs and knowledge gaps

At the start of this research, we conducted a critical review of international literature on the state-of-art knowledge on eutrophication and nutrient releases in urban areas in SSA (Chapter 2). We found that less than 30% of the large cities in sub-Saharan Africa were sewered and about 80% of the wastewater produced in these cities was untreated and was therefore either discharged into the soil via leachates from on-site sanitation systems or directly discharged into urban streams, rivers and lakes. As a consequence, most lakes in large cities in SSA (e.g. Lake Victoria, Lake McIlwaine etc.) were eutrophic due to the excessive discharge of wastewater-derived nutrients (N and P) into the environment. We attributed this condition to the rapid increase in population and urbanization in cities, especially in informal settlements (or slums), where there is uncontrolled disposal of wastewater. Existing literature showed that there is a large knowledge gap regarding fate and transport of these nutrients in groundwater and surface water in unsewered urban areas. More specifically, there were no insights into the main sources of nutrients and the hydro-chemical and geochemical processes that regulate or influence the transport of nutrients (N and P) and other contaminants (e.g. micro-pollutants) in surface water and groundwater. These gaps formed the basis of our research and were systematically addressed as summarized below.

Source of nutrients in the shallow groundwater in unsewered urban slum catchments

We used hydro-chemical tracers (i.e. major cations and anions) to track the main sources of nutrients (N and P) in an unsewered slum-dominated catchment in Kampala Uganda (i.e. the Lubigi catchment; chapter 3). The studied catchment, like in many parts of SSA, is underlain by Precambrian basement rocks consisting of granite-gneiss rocks overlain by deeply weathered and poorly-buffering lateritic regolith soils. The fractured rocks and the weathered regolith are productive (conductivities of up to 3 m/d) and are important aquifers of the shallow groundwater flow systems that normally discharge in valley springs. The shallow groundwater, which we sampled from springs in valleys, was acidic (pH < 5), slightly anoxic and rich in nitrate (NO_3^-), but it had low concentrations of PO_4^{3-} (< 3 μmol/L) and NH_4^+ (average of 0.01 mmol/L). Although the high presence of NO_3^- in groundwater is often attributed to infiltration of wastewater from on-site sanitation (ARGOSS, 2002), hydrochemical evidence in this study showed that atmospheric deposition also contributed to N deposition (in the form of NO_3^- and NH_4^+) to groundwater by precipitation recharge owing to the presence of nitrogen-containing acid rains. Here, we showed that groundwater samples collected from less inhabited parts of the catchment, such as areas with scattered trees and shrubs, had a chemical composition (pH of 4.8, EC of 111 μS/cm, NO_3^- of 0.4 mmol/L or 23 mg/L) that corresponded to the chemical composition of rain water that had been concentrated by evaporation (Chapter 3). The presence of nitrogen-containing rains (or dilute nitric acid) was attributed to the high levels of air pollution (in the form of nitrogen oxides, N_xO and ammonia, NH_3) caused by exhaust from the increasing number of motor vehicles in Kampala city. In parts of the catchment where slum areas were located, hydro-chemical evidence showed that wastewater leachates further influenced the presence of nutrients in groundwater, especially nitrate. Using the statistical hierarchical cluster analysis, we showed that the most polluted springs (NO_3^- of 2 mmol/L and EC of 688 μS/cm) were all located in slum areas whereas less polluted springs (NO_3^- of 0.37 mmol/L and EC of 111 μS/cm) were located in uninhabited areas of the catchment. We therefore concluded that both atmospheric deposition and wastewater leachates from on-site sanitation systems, especially in slum areas, introduced a significant amount of nutrients to groundwater.

In previous studies, wastewater leachates from on-site sanitation systems to shallow groundwater have been attributed to presence of pit latrines and refuse dump sites (Katukiza et al., 2012; Kulabako et al., 2007). However, results from our micro-plot investigations carried out in Bwaise III parish slum, which overlies a low-lying alluvial aquifer (Chapter 4) suggest that refuse dumps sites did not significantly pollute groundwater because the hydro-chemical concentrations downgradient of the dump site were not statistically significantly different ($p > 0.05$) from upgradient concentrations. Instead, most contamination originated from pit latrines with leachate concentrations containing up to 2.4 mg/L (or 26 μmol/L) as PO_4^{3-}, 57 mg/L (or 3.2 mmol/L) as NH_4^+ and 228 mg/L (or 3.7 mmol/L) as NO_3^-. These concentrations are much higher than the eutrophication threshold of 2.4 μmol/L for P and 0.1 mmol/L for N. It is important to note here that Bwaise III parish slum is located in a low-lying alluvial aquifer whose hydro-geologic formation is different from the upper areas of the catchment. Here, the aquifer was very shallow (< 1 m) and the redox environment was anoxic (NO_3^- to Fe reducing; Chapter 4 and Chapter 5) contrary to the regolith aquifer which was oxic. Our results also revealed that pit latrines do a very good job in retaining nutrients. By comparing the N and P loads in groundwater with the potential nutrient mass input from pit latrine usage, we found that about 99% of P and over 80% of N was retained in the pit latrine and the shallow sub-surface in the immediate vicinity of the pit latrines (Chapter 4).

Fate of nutrients in groundwater in urban slum areas and catchments

Wastewater leaching to groundwater contains nutrients predominantly in the form of ammonium (NH_4^+) for N species and ortho-phosphate (PO_4^{3-}) for P species (Correll, 1998; Thornton et al., 1999).The fate of these nutrients in shallow groundwater is summarized here considering two hydro-geological contexts: the regolith aquifer especially in the upper parts of the catchment (see Chapter 3) and the alluvial sandy aquifer in low-lying areas where slums are usually located (see Chapters 4 and 5).

The regolith aquifer: The shallow groundwater in the regolith aquifer we studied was slightly oxic (between NO_3^- and Mn-reducing; DO of 2 mg/L) and the NH_4^+ in the wastewater leachates was rapidly oxidized to NO_3 by nitrification (Chapter 3). Hence, groundwater was rich in nitrate, a relatively conservative species in aerobic groundwater. Consequently, groundwater was acidic (pH < 5) following the release of H^+ ions during nitrification. However, we saw earlier that acid rain recharge also contributed to the acidity of groundwater. Because the regolith aquifer consisted of deeply weathered silicate rocks, the pH was poorly buffered giving rise to acidic conditions. In such acidic environments, PO_4^{3-} in wastewater leachate can be greatly attenuated by adsorption (Reddy et al., 1999). Indeed the concentrations of PO_4^{3-} were very low (2 μmol/L or 0.062 mgP/L) and they were close to the minimum of 0.075 mgP/L or 2.4 μmol/L required to cause eutrophication (Dodds et al., 1998). We concluded that this attenuation was due to the strong adsorption of P to Fe-/Al-oxides because these oxides were present in large amounts owing to the abundance of the laterite in the study area. Geochemical calculations using the PHREEQC code (Parkhurst and Appelo, 1999) also showed that groundwater was near saturation with respect to $MnHPO_4$. Hence, PO_4^{3-} transport was also likely regulated by precipitating with Mn^{2+}.

The alluvial sandy aquifer: The shallow groundwater in the alluvial sandy aquifer in Bwaise slum was, on the contrary, strongly reducing (Fe-reducing) and had neutral pH values and high concentrations of Fe^{2+}, Ca^{2+}, NH_4^+ and HCO_3^- (Chapter 5). This was attributed to the presence of relatively high organic matter (0.6 – 0.8%) originating from sedimentary organic carbon due to deposition of decaying wetland vegetation, the large input of dissolved organic carbon in wastewater leaching from pit latrines and the long residence time in the alluvial aquifer (about 60 years). Hence, the processes governing N and P transport were different from those in the regolith aquifer described above. For example, groundwater flowing from the regolith was rich in NO_3^- (2 mmol/L; Chapter 3) but in the shallow alluvial aquifer, NO_3^- was almost completely removed by denitrification (< 0.2 mmol/L, Chapter 5). The dominant form of N was therefore NH_4^+ (2.2 mmol/L). NH_4^+ was, however, also partially removed by about 30% in the alluvial aquifer because the NH_4^+ concentrations in groundwater were slightly less than the potential input from pit latrines (3.2 mmol/L). We inferred that the process likely taking place here was anaerobic ammonium oxidation (*Anammox*).

As with the oxic regolith aquifer, PO_4^{3-} leaching from pit latrines to the alluvial aquifer was also greatly attenuated by sorption and mineral precipitation. The average PO_4^{3-} concentration in the shallow groundwater beneath Bwaise slum was 6 µmol/L (or 0.18 mgP/L), which is about 75% less than the potential input of 26 µmol/L from pit latrines (chapter 3) but still slightly higher than the minimum of 0.075 mg TP/L or 2.4 µmol/L (Dodds et al., 1998) required to cause eutrophication. The attenuation of PO_4^{3-} in the shallow alluvial aquifer was attributed to adsorption of P to sediment calcite and co-precipitation of P with calcite minerals in groundwater. This conclusion was supported by the following facts: (1) the maximum adsorption capacity, C_{max}, of P onto Fe-oxides in the aquifer was low (C_{max} < 2 mg/kg) compared to the available P (8 – 18 mg/kg) (2) groundwater was saturated with respect to calcite, (3) the actual C_{max} of the sediments ($\cong 600$ mg/kg) was statistically correlated with Ca, and (4) the presence of high concentrations of Ca and HCO_3^- in the alluvial aquifer. In general, our results showed that these aquifers appeared to remove a substantial amount of nutrients ($> 75\%$ for PO_4^{3-}, 100% for NO_3^- and $> 30\%$ of NH_4^+) that were potentially leaching from pit latrines and other polluting on-site sanitation systems. A schematic representation of the fate of nutrients in the shallow alluvial aquifers beneath the studied slum area and catchment is presented in Fig. 7.1.

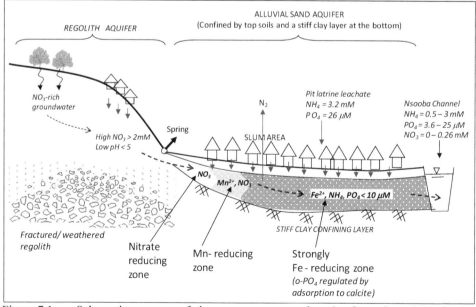

Figure 7.1:   Schematic concept of the processes governing the fate of nutrients in the alluvial sandy aquifer underlying the studied slum area.

Source and fate of nutrients in surface water

The primary channel draining the studied catchment (Lubigi) contained very high concentrations of nutrients in the form of PO_4^{3-} and NH_4^+. The base flow concentrations of ortho-phosphates at the outlet of the studied catchment ranged from 0.11 to 0.78 mgP/L (average of 0.36 mgP/L), total phosphorus ranged from 0.51 to 1.61 mgP/L (average of 1.15 mgP/L) and ammonium ranged from 6.8 to 12.1 mgN/L (average of 10 mgN/L as NH_3-N). The P concentrations were about 16 times more than the eutrophication limit of 0.075 mgP/L (Dodds et al., 1998). Consequently, the resulting P fluxes (0.3 kg km^2 d^{-1} for PO_4-P and 0.95 kg km^2 d^{-1} for TP) were much higher than values normally reported in published literature for other catchments (see chapter 6). The results further showed that surface water was anoxic (Mn-reducing) and as a consequence the NO_3^- concentrations were very low. We therefore inferred that the NO_3-rich groundwater in the regolith aquifer did not contribute to nutrients (N and P) in surface water because NO_3^- was likely lost by denitrification upon groundwater exfiltration whereas ortho-phosphates were already attenuated by adsorption and mineral precipitation in the regolith aquifer (see chapter 3). With these revelations, we concluded that the most important source of nutrients measured in surface water was the direct discharge of domestic wastewater into drainage channels, and particularly from grey water. Grey water is the main wastewater stream in the studied channels (Katukiza et al., 2012, 2014) and it comprises of untreated wastewater derived from bathrooms, kitchens and laundry. This wastewater stream normally ends up into the primary channels via tertiary channels that drain individual households, especially in slum areas. Our results revealed that the P transported in the primary channel during low flows was regulated by 1) the co-precipitation of P with calcite, 2) the adsorption of P to Fe-oxides, and 3) the settling of organic matter. Therefore, a significant amount of P was also retained in the sediments via these processes. However, the P retained in the sediments also seemed to be an important P pool that contributed to internal P loading in drainage channels. This is because, the channel bed sediments were P saturated and showed tendencies to release P by desorption of Fe-bound P and mineralization of organic bound P in the sediments (Chapter 6). During storm runoff events, it was clear that there was a flushing of P to downstream areas owing to the resuspension of the sediments that had accumulated along the channel beds. These results indicated that urban catchments with informal settlements posed high threats to downstream ecosystems due to large fluxes of P derived from both external and internal P loadings as well as flushing effects.

Implications and recommendations

An important scientific contribution of this study is that in addition to wastewater leaching from on-site sanitation, atmospheric deposition can also be an important source of nutrients (particularly nitrate) found in groundwater in growing cities in sub-Saharan Africa. Nutrient pollution management strategies should therefore consider an integrated approach and also focus on limiting air pollution, particularly from motor vehicle exhausts and the combustion of fossil fuels. Such an integrated approach requires that researchers collaborate with authorities and relevant stakeholders such that their views and roles are considered in the overall urban water management and governance (Raadgever et al., 2012).

Another important contribution of this research is that a combination of pit latrines and underlying aquifers (both regolith and alluvial) provide a natural and effective attenuation of nutrients leaching into groundwater. Polluted groundwater may therefore not be of major concern regarding eutrophication threats to surface water. Hence, promoting the use and maintenance of on-site sanitation systems such as pit latrines or equivalently septic tanks, may in fact be sufficient to reduce nutrient pollution to groundwater and in a long run contribute to improvement of the water quality of downstream ecosystems.

Contrary to groundwater, surface water in drainage channels seemed to pose a serious threat to eutrophication of downstream water resources. In the studied urban catchment, we measured high concentrations of sanitation-related nutrients (particularly, PO_4^{3-} and NH_4^+) in the Nsooba channel, which drained the catchment. These nutrient concentrations were about 16 times the minimum required to impair the water quality of fresh water bodies by eutrophication. Nutrient management strategies in unsewered urban poor areas should therefore place more emphasis on controlling sanitation-related nutrients exported via surface water, with focus on P because it is considered to be the limiting nutrient for eutrophication. One way of doing this is to try to have a complete understanding of the processes regulating the transport of P in these channels. Our research for example provided insights into the main mechanisms regulating phosphorus transported in these drainage channels. We found that P was regulated by the 1) adsorption of P to calcite precipitates during base flows, 2) adsorption of P to Fe-oxides during and after rain events when there is re-suspension of Fe-rich bed sediments and, 3) settling of organic P. These mechanisms contributed to the retention of P to bed sediments. During high flows, the P retained in the bed sediments was occasionally flushed out of the system to downstream ecosystems. Our study was, however, not conclusive on the presence of first-flush effects. The processes identified could be useful in developing process-based hydrological models as well as water-quality models for predicting and quantifying phosphorus loads from these types of catchments. Results from such models can be useful for policy and decision making on the strategies required to reduce the nutrient (N and P) exported from urban catchments.

Although process-based models could be useful in developing strategies to reduce nutrient exports from urban catchments, a simple phosphorus management practice could focus on reducing phosphorus loads at the points of generation (i.e. wastewater effluents). In this study, the discharge of untreated grey water (wastewater from bathrooms, kitchens and laundry) was identified as the main source of phosphorus to surface water in urban catchments (chapter 6). Treating grey water at individual households, especially in slum areas, could therefore help to reduce the phosphorus exported from urban catchments with informal settlements. Experiments carried out by Katukiza (2014) have, for example, shown that it is possible to effectively treat grey water at household level using a low-cost two-step crushed lava rock. The required treatment efficiency to reduce eutrophication problems can roughly be estimated using the chloride (a conservative tracer) and phosphorus mass balance approaches. Here, we assume that the background mass load in the primary channel during low flows (assumed to be exfiltrated groundwater) and the mass load of the grey water feed from the tertiary channel equals the resultant mass load at the catchment outlet during low flows.

1. Using the chloride mass balance, we can estimate the contribution of grey water to the total discharge (0.22 m^3/s; chapter 3) at the catchment outlet. Using the background Cl$^-$ concentration of 26 mg/L (average of exfiltrating groundwater, Chapter 3), a Cl$^-$ concentration of 78 mg/L in the grey water feed (Chapter 6) and the resultant Cl$^-$ concentration of 40 mg/L at the catchment outlet (chapter 6), then by mass balance, grey water contributes approximately 25% ($= 0.05$ m^3/s) of the total base flow.

2. Using the TP mass balance, we can then estimate the required grey water treatment efficiency. Using the background TP concentration of 0.06 mgP/L (average of groundwater, Chapter 3) and the TP concentration of 5.2 mgP/L in the grey water feed (Chapter 6), then to achieve the required TP concentration of 0.075 mg/L (eutrophication

limit; Dodds et al., 1998) at the catchment outlet, the required treatment efficiency would have to be 97 %.

To reduce nutrient pollution from urban informal catchments to downstream water resources to acceptable levels, a good starting point may be therefore to install grey water treatment units at household level ensuring a treatment efficiency of at least 95% for phosphorus. For such an intervention to work in slum areas, there is a need to consider an integrated approach that takes into account the various social and economic aspects that may hinder greater participation of the beneficiaries and hence sustainability of these interventions. Some of these aspects include installation and operation costs, land required for installation, ease of operation and maintenance, cultural considerations and so on. A promising approach that has been successively applied in high density informal settlements in SSA (e.g. Uganda, Kenya, Ethiopia and South Africa) is the use of grey water tower gardens (e.g. Kulabako et al., 2011; Shewa and Geleta, 2010). These grey water towers provide a simple and low-cost method to treat and reuse grey water generated at households. They consist of simple circular plastic bags filled with soil, ash and compost mixture, and a gravel column at the centre. The grey water is poured in the bag through the gravel column and the treated water is reused to grow vegetables, which are usually planted in holes cut at the sides of the bag (see Kulabako et al., 2011). Such low-cost treatment systems that provide incentives to the users (e.g. growing crops) could be important avenues to treat grey water and reduce the nutrients discharged from slum areas into the environment.

7.2. Future research needs

Obviously, more research is needed to properly understand the fate and transport of nutrients in groundwater and surface water in urban informal settlements. Here, we identify potential avenues for future research which were not addressed in this study but build on the existing results:

- Comparative studies should be carried out in other informal settlements in sub-Saharan Africa to relate and contrast the results. This will help to assess how far the findings can be regionalized to other similar environments. Emphasis should be placed on understanding the dynamics and hydrochemistry of episodic flood drainage waters in order to seek ways of reducing the amount of phosphorus exported via surface water from these types of catchments to downstream ecosystems.

- Environmental isotopes such as $\delta^{15}N$ and $\delta^{18}O$ can be used to differentiate among the various sources and pathways of nutrients (especially NO_3^-) found in shallow groundwater. This would also enable gaining further insights into the actual residence times of water and nutrients in the various hydrological systems (unsaturated zone, shallow/deep groundwater, surface water etc.) (Nyarko et al., 2010). These tracers have not been applied in contaminated waters in unsewered urban informal settlements of sub-Saharan Africa, but are promising tools to (partly) unfold the complexity in these systems.

- Artificial DNA tracers, which can be detected using the real time quantitative polymerase chain reaction (qPCR), are also emerging as important and alternative tracers for understanding hydrological and geochemical processes (Foppen et al., 2013). Some of the dominant processes identified in this study (e.g. nitrification, denitrification and anammox) can also be demonstrated by quantifying the DNA of microbial communities responsible for these processes using the qPCR.

- The adsorption of P to the aquifer material and mineral precipitation are the two main processes regulating the sub-surface transport of ortho-phosphate. In our study, we did not determine the relative contribution of these two processes to the retention of P. In addition, the sorption capacity of the aquifer can become depleted but it is not clear for which conditions phosphate can be re-mobilized especially in urban slums areas. In addition, we recommend that future studies carry out high-resolution monitoring of phosphorus forms and other chemical parameters during base flow and high flow events in order to have a complete understanding of the transport of P during different hydrological situations. This could be possible by using automatic samplers and field-automated analyzing equipment because the manual analyzing procedure is difficult and too time consuming.

- More research is needed to develop process-based models that can effectively simulate the transport and transformations of sanitation-related nutrients in groundwater and surface water in slum-dominated catchments and to investigate the impacts of alternative sanitation systems on nutrient discharge to downstream ecosystems.

- Future research should also focus on understanding and modeling the transport and fate of emerging micro-pollutants such as pharmaceuticals and estrogens of personal care products in the hydrologic cycle in urban slum catchments and beyond. These micro-pollutants are often carcinogenic and with the increasing urbanization, they could be widespread in the local/regional hydrological systems of sub-Saharan Africa.

References

Abiye, T., Sulieman, H., and Ayalew, M., 2009, Use of treated wastewater for managed aquifer recharge in highly populated urban centers: a case study in Addis Ababa, Ethiopia: Environmental Geology, v. 58, p. 55-59.

Adeniji, H.A., Denny, P., and Toerien, D.F., 1981, Man-made lakes, *in* Symoens, J.J., Burgis, M., and Gaudet, J.J., eds., The Ecology and Utilization of African Inland Waters, p. 125–34.

Alagbe, S.A., 2006, Preliminary evaluation of hydrochemistry of the Kalambaina Formation, Sokoto Basin, Nigeria: Environmental Geology, v. 51, p. 39-45.

Allanson, B.R., and Gieskes, J.M.T.M., 1961, An introduction to the limnology of Hartbeestpoortdam with special reference to the effect of industrial and domestic pollution: Hydrobiologia, v. 18, p. 77-94.

Andrade, A., and Stigter, T., 2011, Hydrogeochemical controls on shallow alluvial groundwater under agricultural land: case study in Central Portugal: Environmental Earth Sciences, v. 63, p. 809-825.

APHA/AWWA/WEF, 2005, Standard Methods for the Examination of Water and Wastewater, 21st ed: Washington, DC., American Public Health Association.

Appelo, C.A.J., and Postma, D., 2007, Geochemistry, groundwater and pollution: Leiden, The Netherlands, A.A Balkema Publishers, 558 p.

ARGOSS, 2002, Assessing Risk to Groundwater from On-site Sanitation: Scientific Review and Case Studies, Volume Commissioned Report, British Geological Survey, p. 44.

Arimoro, F.O., Ikomi, R.B., and Iwegbue, C.M.A., 2007, Water quality changes in relation to Diptera community patterns and diversity measured at an organic effluent impacted stream in the Niger Delta, Nigeria: Ecological Indicators, v. 7, p. 541-552.

Barel, C.D.N., Dorit, R., Greenwood, P.H., Fryer, G., Hughes, N., Jackson, P.B.N., Kawanabe, H., Lowe-McConnell, R.H., Nagoshi, M., Ribbink, A.J., Trewavas, E., Witte, F., and Yamaoka, K., 1985, Destruction of fisheries in Africa's lakes: Nature, v. 315, p. 19-20.

Barrett, M., Nalubega, M., and Pedley, S., 1999a, On-site sanitation and urban aquifer systems in Uganda: Waterlines, v. Vol. 17, No. 4, p. 10-13.

Barrett, M.H., Hiscock, K.M., Pedley, S., Lerner, D.N., Tellam, J.H., and French, M.J., 1999b, Marker species for identifying urban groundwater recharge sources: A review and case study in Nottingham, UK: Water Research, v. 33, p. 3083-3097.

Bedore, P.D., David, M.B., and Stucki, J.W., 2008, Mechanisms of Phosphorus Control in Urban Streams Receiving Sewage Effluent: Water, Air, and Soil Pollution, v. 191, p. 217-229.

Belkhiri, L., Boudoukha, A., Mouni, L., and Baouz, T., 2010, Application of multivariate statistical methods and inverse geochemical modeling for characterization of groundwater—A case study: Ain Azel plain (Algeria): Geoderma, v. 159, p. 390-398.

Bere, T., 2007, The assessment of nutrient loading and retention in the upper segment of the Chinyika River, Harare: Implications for eutrophication control: Water SA, v. 33, p. 279-284.

Beyene, A., Legesse, W., Triest, L., and Kloos, H., 2009, Urban impact on ecological integrity of nearby rivers in developing countries: the Borkena River in highland Ethiopia: Environmental Monitoring and Assessment, v. 153, p. 461-476.

Biryabareema, M., 2001, Groundwater contamination and role of geologic strata in retaining pollutants around waste disposal sites in Greater Kampala. PhD Thesis: Makerere University, Kampala, Uganda.

Blanco, A.C., Nadaoka, K., Yamamoto, T., and Kinjo, K., 2010, Dynamic evolution of nutrient discharge under stormflow and baseflow conditions in a coastal agricultural watershed in Ishigaki Island, Okinawa, Japan: Hydrological Processes, v. 24, p. 2601-2616.

Boers, P., and de Bles, F., 1991, Ion concentrations in interstitial water as indicators for phosphorus release processes and reactions: Water Research, v. 25, p. 591-598.

Borggaard, O., 1979, Selective extraction of amorphous iron oxides by EDTA from a Danish sandy loam: Journal of Soil Science, v. 30, p. 727-734.

Borggaard, O.K., Jdrgensen, S.S., Moberg, J.P., and Raben-Lange, B., 1990, Influence of organic matter on phosphate adsorption by aluminium and iron oxides in sandy soils: Journal of Soil Science, v. 41, p. 443-449.

Bouwer, H., 1989, The Bouwer and Rice slug test - an update: Ground Water, v. 27, p. 304-309.

Bouyoucos, G.J., 1962, Hydrometer method improved for making particle size analysis of soils.: Agronomy Journal, v. 53, p. 464-465.

Boyle, F., and Lindsay, W., 1986, Manganese phosphate equilibrium relationships in soils: Soil Science Society of America Journal, v. 50, p. 588-593.

Bray, R.H., and Kurtz, L.T., 1945, Determination of total, organic, and available forms of phosphorus in soils: Soil Sci., v. 59, p. 39-45.

Brdjanovic, D., Slamet, A., Van Loosdrecht, M.C.M., Hooijmans, C.M., Alaerts, G.J., and Heijnen, J.J., 1998, Impact of excessive aeration on biological phosphorus removal from wastewater: Water Research, v. 32, p. 200-208.

Brender, J.D., Weyer, P.J., Romitti, P.A., Mohanty, B.P., Shinde, M.U., Vuong, A.M., Sharkey, J.R., Dwivedi, D., Horel, S.A., and Kantamneni, J., 2013, Prenatal nitrate intake from drinking water and selected birth defects in offspring of participants in the National Birth Defects Prevention Study: Environmental Health Perspectives, v. 121, p. 1083 - 1089.

Butler, A.P., Brook, C., Godley, A., Lewin, K., and Young, C.P., 2003a, Attenuation of landfill leachate in unsaturated sandstone, *in* Christensen, T.H., Cossu, R., and Stegmann, R., eds., Proceedings Sardinia 2003, 9th International Landfill Symposium., CISA Publisher.

Butler, J.J., Jr., Garnett, E.J., and Healey, J.M., 2003b, Analysis of slug tests in formations of high hydraulic conductivity: Ground Water, v. 41, p. 620-30.

Byamukama, D., Kansiime, F., Mach, R.L., and Farnleitner, A.H., 2000, Determination of Escherichia coli contamination with chromocult coliform agar showed a high level of discrimination efficiency for differing fecal pollution levels in tropical waters of Kampala, Uganda: Applied and Environmental Microbiology, v. 66, p. 864-868.

Campbell, L.M., Wandera, S.B., Thacker, R.J., Dixon, D.G., and Hecky, R.E., 2005, Trophic niche segregation in the Nilotic ichthyofauna of Lake Albert (Uganda, Africa): Environmental Biology of Fishes, v. 74, p. 247-260.

Cao, X., Harris, W.G., Josan, M.S., and Nair, V.D., 2007, Inhibition of calcium phosphate precipitation under environmentally-relevant conditions: Science of the Total Environment, v. 383, p. 205-215.

Capell, R., Tetzlaff, D., Malcolm, I.A., Hartley, A.J., and Soulsby, C., 2011, Using hydrochemical tracers to conceptualise hydrological function in a larger scale catchment draining contrasting geologic provinces: Journal of Hydrology, v. 408, p. 164-177.

Carlyle, G.C., and Hill, A.R., 2001, Groundwater phosphate dynamics in a river riparian zone: effects of hydrologic flowpaths, lithology and redox chemistry: Journal of Hydrology, v. 247, p. 151-168.

Cash, D.W., and Moser, S.C., 2000, Linking global and local scales: designing dynamic assessment and management processes: Global Environmental Change, v. 10, p. 109-120.

Christensen, T.H., Kjeldsen, P., Albrechtsen, H.J.r., Heron, G., Nielsen, P.H., Bjerg, P.L., and Holm, P.E., 1994, Attenuation of landfill leachate pollutants in aquifers: Critical Reviews in Environmental Science and Technology, v. 24, p. 119-202.

Chua, L.H.C., Lo, E.Y.M., Shuy, E.B., and Tan, S.B.K., 2009, Nutrients and suspended solids in dry weather and storm flows from a tropical catchment with various proportions of rural and urban land use: Journal of Environmental Management, v. 90, p. 3635-3642.

CIDI, 2006, Citizens action for water and sanitation survey report in Kawempe division, Kampala, Uganda: Community Integrated Development Initiatives (CIDI).

Clark, T., Stephenson, T., and Pearce, P., 1997, Phosphorus removal by chemical precipitation in a biological aerated filter: Water Research, v. 31, p. 2557-2563.

Cloutier, V., Lefebvre, R., Therrien, R., and Savard, M.M., 2008, Multivariate statistical analysis of geochemical data as indicative of the hydrogeochemical evolution of groundwater in a sedimentary rock aquifer system: Journal of Hydrology, v. 353, p. 294-313.

Coetzee, M.A.A., Van der Merwe, M.P.R., and Badenhorst, J., 2011, Effect of nitrogen loading rates on nitrogen removal by using a biological filter proposed for ventilated improved pit latrines: Int. J. Environ. Sci. Tech, v. 8, p. 363-372.

Cooke, J.G., Stub, L., and Mora, N., 1992, Fractionation of Phosphorus in the Sediment of a Wetland after a Decade of Receiving Sewage Effluent: J. Environ. Qual., v. 21, p. 726-732.

Corbett, D.R., Kump, L., Dillon, K., Burnett, W., and Chanton, J., 2000, Fate of wastewater-borne nutrients under low discharge conditions in the subsurface of the Florida Keys, USA: Marine Chemistry, v. 69, p. 99-115.

Correll, D.L., 1998, The Role of Phosphorus in the Eutrophication of Receiving Waters: A Review: J. Environ. Qual., v. 27, p. 261-266.

Costello, A.B., and Osborne, J.W., 2005, Best practices in exploratory factor analysis: four recommendations for getting the most from your analysis: Practical Assessment, Research & Evaluation, v. 10.

Coulter, G.W., and Jackson, P.N.B., 1981, Deep Lakes, *in* Symoens, J.J., Burgis, M., and Gaudet, J.J., eds., The ecology and utilization of African inland waters, p. 114–24.

Cózar, A., Bergamino, N., Mazzuoli, S., Azza, N., Bracchini, L., Dattilo, A.M., and Loiselle, S.A., 2007, Relationships between wetland ecotones and inshore water quality in the Ugandan coast of Lake Victoria: Wetlands Ecology and Management, v. 15, p. 499-507.

Cronin, A.A., Hoadley, A.W., Gibson, J., Breslin, N., Komou, F.K., Haldin, L., and Pedley, S., 2007, Urbanisation effects on groundwater chemical quality: findings focusing on the nitrate problem from 2 African cities reliant on on-site sanitation: Journal of Water and Health, v. 5, p. 441-454.

Cronin, A.A., Pedley, S., Okotto-Okotto, J., Oginga, J.O., and Chenoweth, J., 2006, Degradation of groundwater resources under African cities: Technical and socio-economic challenges, *in* Xu, Y., and Usher, B., eds., Groundwater pollution in Africa: Leiden, The Netherlands, Taylor & Francis/Balkema, p. 89-97.

Cronin, A.A., Taylor, R.G., Powell, K.L., Barrett, M.H., Trowsdale, S.A., and Lerner, D.N., 2003, Temporal variations in the depth-specific hydrochemistry and sewage-related microbiology of an urban sandstone aquifer, Nottingham, United Kingdom: Hydrogeology Journal, v. 11, p. 205-216.

Das, S.K., Routh, J., Roychoudhury, A.N., Klump, J., and Ranjan, R.K., 2009, Phosphorus dynamics in shallow eutrophic lakes: an example from Zeekoevlei, South Africa: Hydrobiologia, v. 619, p. 55-66.

Das, S.K., Routh, J., Roychoudhury, A.N., and Klump, J.V., 2008, Elemental (C, N, H and P) and stable isotope (delta N-15 and delta C-13) signatures in sediments from Zeekoevlei, South Africa: a record of human intervention in the lake: Journal of Paleolimnology, v. 39, p. 349-360.

Datry, I., Malard, F., and Gibert, J., 2004, Dynamics of solutes and dissolved oxygen in shallow urban groundwater below a stormwater infiltration basin: Science of the Total Environment, v. 329, p. 215-229.

Davies, T.C., and Mundalamo, H.R., 2010, Environmental health impacts of dispersed mineralisation in South Africa: Journal of African Earth Sciences, v. 58, p. 652-666.

de Mesquita Filho, M., and Torrent, J., 1993, Phosphate sorption as related to mineralogy of a hydrosequence of soils from the Cerrado region (Brazil): Geoderma, v. 58, p. 107-123.

De Villiers, G., and Malan, E., 1985, The water quality of a small urban catchment near Durban, South Africa: Water SA, v. 11, p. 35-40.

De Villiers, S., 2007, The deteriorating nutrient status of the Berg River, South Africa: Water SA, v. 33, p. 659-664.

Deletic, A., 1998, The first flush load of urban surface runoff: Water Research, v. 32, p. 2462-2470.

Deutsch, W.J., 1997, Groundwater Chemistry - Fundamentals and Applications to Contamination: New York, Lewis Publishers, 221 p.

Devi, R., Tesfahune, E., Legesse, W., Deboch, B., and Beyene, A., 2008, Assessment of siltation and nutrient enrichment of Gilgel Gibe dam, Southwest Ethiopia: Bioresource Technology, v. 99, p. 975-979.

Diaz, O., Reddy, K., and Moore, P., 1994, Solubility of inorganic phosphorus in stream water as influenced by pH and calcium concentration: Water Research, v. 28, p. 1755-1763.

Didszun, J., and Uhlenbrook, S., 2008, Scaling of dominant runoff generation processes: Nested catchments approach using multiple tracers: Water Resources Research, v. 44.

Dillion, P., 1997, Groundwater pollution by sanitation on tropical islands: Adelaide, Australia, International Hydrological Programme - IHP-V Project 6-1, UNESCO.

Diop, S., and Tijani, M.N., 2008, Assessing the basement aquifers of Eastern Senegal: Hydrogeology Journal, v. 16, p. 1349-1369.

Dodds, W.K., Jones, J.R., and Welch, E.B., 1998, Suggested classification of stream trophic state: distributions of temperate stream types by chlorophyll, total nitrogen, and phosphorus: Water Research, v. 32, p. 1455-1462.

Douglas, I., Alam, K., Maghenda, M., Mcdonnell, Y., McLean, L., and Campbell, J., 2008, Unjust waters: climate change, flooding and the urban poor in Africa: Environment and Urbanization, v. 20, p. 187-205.

Dzombak, D.A., and Morel, F.M., 1990, Surface complexation modeling: hydrous ferric oxide, Wiley New York.

Dzwairo, B., Hoko, Z., Love, D., and Guzha, E., 2006, Assessment of the impacts of pit latrines on groundwater quality in rural areas: A case study from Marondera district, Zimbabwe: Physics and Chemistry of the Earth, v. 31, p. 779-788.

Edet, A., and Okereke, C., 2005, Hydrogeological and hydrochemical character of the regolith aquifer, northern Obudu Plateau, southern Nigeria: Hydrogeology Journal, v. 13, p. 391-415.

Efe, S.I., 2005, Quality of water from hand dug wells in Onitsha Metropolitan areas of Nigeria.: The Environmentalist, v. 25, p. 5-12.

Enabor, B., 1998, Integrated Water Management by Urban Poor Women: A Nigerian Slum Experience: International Journal of Water Resources Development, v. 14, p. 505-512.

Evans, D.J., Johnes, P.J., and Lawrence, D.S., 2004, Physico-chemical controls on phosphorus cycling in two lowland streams. Part 2–the sediment phase: Science of the Total Environment, v. 329, p. 165-182.

Faillat, J.P., 1990, Sources of Nitrates in Fissure Groundwater in the Humid Tropical Zone - the Example of Ivory-Coast: Journal of Hydrology, v. 113, p. 231-264.

FAO, 1970, Evaporation in East Africa. Bulletin of International Association of Scientific hydrology XVI 3/1970, Food and Agricultural Organisation.

Fatoki, O.S., Muyima, N.Y.O., and Lujiza, N., 2001, Situation analysis of water quality in the Umtata River Catchment: Water SA, v. 27, p. 467-474.

Feigin, A., Ravina, I., and Shalhevet, J., 1991, Irrigation with Treated Sewage Effluent: Management for Environmental Protection.: Springer, Berlin, v. ISBN 3-540-50804-X.

Fenton, O., Richards, K.G., Kirwan, L., Khalil, M.I., and Healy, M.G., 2009, Factors affecting nitrate distribution in shallow groundwater under a beef farm in South Eastern Ireland: Journal of Environmental Management, v. 90, p. 3135-3146.

Figueredo, C.C., and Giani, A., 2001, Seasonal variation in the diversity and species richness of phytoplankton in a tropical eutrophic reservoir: Hydrobiologia, v. 445, p. 165-174.

Flynn, R., Taylor, R., Kulabako, R., and Miret-Gaspa, M., 2012, Haematite in Lateritic Soils Aids Groundwater Disinfection: Water, Air, and Soil Pollution, v. 223, p. 2405-2416.

Foppen, J.W., Seopa, J., Bakobie, N., and Bogaard, T., 2013, Development of a methodology for the application of synthetic DNA in stream tracer injection experiments: Water resources research, v. 49, p. 5369-5380.

Foppen, J.W.A., 2002, Impact of high-strength wastewater infiltration on groundwater quality and drinking water supply: the case of Sana'a, Yemen: Journal of Hydrology, v. 263, p. 198-216.

Foppen, J.W.A., and Kansiime, F., 2009, SCUSA: integrated approaches and strategies to address the sanitation crisis in unsewered slum areas in African mega-cities: Reviews in Environmental Science and Bio/Technology, v. 8, p. 305-311.

Foppen, J.W.A., van Herwerden, M., Kebtie, M., Noman, A., Schijven, J.F., Stuyfzand, P.J., and Uhlenbrook, S., 2008, Transport of Escherichia coli and solutes during waste water infiltration in an urban alluvial aquifer: Journal of Contaminant Hydrology, v. 95, p. 1-16.

Fox, L.E., Sager, S.L., and Wofsy, S.C., 1986, The chemical control of soluble phosphorus in the Amazon estuary: Geochimica et Cosmochimica Acta, v. 50, p. 783-794.

Franceys, R., Pickford, J., and Reed, R., 1992, A guide to the development of on-site sanitation: Geneva, WHO.

Freeman, J.S., and Rowell, D.L., 1981, The adsorption and precipitation of phosphate onto calcite: Journal of Soil Science, v. 32, p. 75-84.

Froelich, P.N., 1988, Kinetic Control of Dissolved Phosphate in Natural Rivers and Estuaries: A Primer on the Phosphate Buffer Mechanism: Limnology and Oceanography, v. 33, p. 649-668.

Gelinas, Y., Randall, H., Robidoux, L., and Schmit, J.P., 1996, Well water survey in two districts of Conakry (Republic of Guinea), and comparison with the piped city water: Water Research, v. 30, p. 2017-2026.

Golterman, H.L., 1995, The role of the ironhydroxide-phosphate-sulphide system in the phosphate exchange between sediments and overlying water: Hydrobiologia, v. 297, p. 43-54.

Golterman, H.L., and De Oude, N.T., 1991, Eutrophication of lakes, rivers and coastal seas., *in* Hutzinger, O., ed., The Handbook of Environmental Chemistry, Volume 5 (Part A): Berlin, Springer-Verlag, p. 79-124.

Golterman, H.L., and Meyer, M.L., 1985, The geochemistry of two hard water rivers, the Rhine and the Rhone, Part 4: The determination of the solubility product of hydroxy-apatite: Hydrobiologia, v. 126, p. 25-29.

Graham, J.P., and Polizzotto, M.L., 2013, Pit Latrines and Their Impacts on Groundwater Quality: A Systematic Review: Environmental Health Perspectives, v. 121, p. 521.

Griffioen, J., 2006, Extent of immobilisation of phosphate during aeration of nutrient-rich, anoxic groundwater: Journal of Hydrology, v. 320, p. 359-369.

Güler, C., Kurt, M.A., Alpaslan, M., and Akbulut, C., 2012, Assessment of the impact of anthropogenic activities on the groundwater hydrology and chemistry in Tarsus coastal plain (Mersin, SE Turkey) using fuzzy clustering, multivariate statistics and GIS techniques: Journal of Hydrology, v. 414, p. 435-451.

Güler, C., and Thyne, G.D., 2004, Hydrologic and geologic factors controlling surface and groundwater chemistry in Indian Wells-Owens Valley area, southeastern California, USA: Journal of Hydrology, v. 285, p. 177-198.

Güler, C., Thyne, G.D., McCray, J.E., and Turner, A.K., 2002, Evaluation of graphical and multivariate statistical methods for classification of water chemistry data: Hydrogeology Journal, v. 10, p. 455-474.

Gunnars, A., Blomqvist, S., Johansson, P., and Andersson, C., 2002, Formation of Fe (III) oxyhydroxide colloids in freshwater and brackish seawater, with incorporation of phosphate and calcium: Geochimica et Cosmochimica Acta, v. 66, p. 745-758.

Haruna, R., Ejobi, F., and Kabagambe, E.K., 2005, The quality of water from protected springs in Katwe and Kisenyi parishes, Kampala city, Uganda: African health sciences, v. 5, p. 14-20.

Hecky, R.E., 1993, The eutrophication of Lake Victoria: Verh Internat Verein Limnol, v. 25, p. 39-48.

Hecky, R.E., and Bugenyi, F.W.B., 1992, Hydrology and chemistry of the African Great Lakes and water issues: problems and solutions.: Mitt. Int. Ver. Limnol., v. 23, p. 45-54.

Hecky, R.E., Bugenyi, F.W.B., Ochumba, P., Talling, J.F., Mugidde, R., Gophen, M., and Kaufman, L., 1994, Deoxygenation of the deep-water of Lake Victoria, East Africa: Limnology and Oceanography, v. 39, p. 1476-1481.

Hinsby, K., Bjerg, P.L., Andersen, L.J., Skov, B., and Clausen, E.V., 1992, A Mini slug test method for determination of a local hydraulic conductivity of an unconfined sandy aquifer: Journal of Hydrology, v. 136, p. 87-106.

Hiscock, K.M., Lloyd, J.W., and Lerner, D.N., 1991, Review of Natural and Artificial Denitrification of Groundwater: Water Research, v. 25, p. 1099-1111.

Hoffmann, C.C., Berg, P., Dahl, M., Larsen, S.E., Andersen, H.E., and Andersen, B., 2006, Groundwater flow and transport of nutrients through a riparian meadow–field data and modelling: Journal of Hydrology, v. 331, p. 315-335.

Hooda, P., Rendell, A., Edwards, A., Withers, P., Aitken, M., and Truesdale, V., 2000, Relating soil phosphorus indices to potential phosphorus release to water: Journal of Environmental Quality, v. 29, p. 1166-1171.

Howard-Williams, C., and Ganf, G.G., 1981, Shallow Waters, *in* Symoens, J.J., Burgis, M., and Gaudet, J.J., eds., The ecology and utilization of African inland waters, p. 103–13.

Howard, G., Pedley, S., Barrett, M., Nalubega, M., and Johal, K., 2003, Risk factors contributing to microbiological contamination of shallow groundwater in Kampala, Uganda: Water Research, v. 37, p. 3421-3429.

Howard, K.W.F., and Karundu, J., 1992, Constraints on the exploitation of basement aquifers in East Africa — water balance implications and the role of the regolith: Journal of Hydrology, v. 139, p. 183-196.

Ikem, A., Osibanjo, O., Sridhar, M.K.C., and Sobande, A., 2002, Evaluation of groundwater quality characteristics near two waste sites in Ibadan and Lagos, Nigeria: Water, Air, and Soil Pollution, v. 140, p. 307-333.

Irwin, J.G., and Williams, M.L., 1988, Acid rain: chemistry and transport: Environ Pollut, v. 50, p. 29-59.

Isunju, J.B., Schwartz, K., Schouten, M.A., Johnson, W.P., and van Dijk, M.P., 2011, Socio-economic aspects of improved sanitation in slums: A review: Public Health, v. 125, p. 368-376.

Jacks, G., Sefe, F., Carling, M., Hammar, M., and Letsamao, P., 1999, Tentative nitrogen budget for pit latrines â€" eastern Botswana: Environmental Geology, v. 38, p. 199-203.

Janardhanan, L., and Daroub, S.H., 2010, Phosphorus Sorption in Organic Soils in South Florida: Soil Science Society of America Journal, v. 74, p. 1597-1606.

Jarvie, H.P., Neal, C., and Withers, P.J., 2006, Sewage-effluent phosphorus: a greater risk to river eutrophication than agricultural phosphorus?: Science of the Total Environment, v. 360, p. 246-53.

Jarvie, H.P., Neal, C., Withers, P.J.A., Wescott, C., and Acornley, R.A., 2005, Nutrient hydrochemistry for a groundwater-dominated catchment: The Hampshire Avon, UK: Science of the Total Environment, v. 344, p. 143-158.

Jarvis, M.J.F., Mitchell, D.S., and Thornton, J.A., 1982, Lake McIlwaine - The Eutrophication and Recovery of a Tropical Man-Made Lake.: The Hague, the Netherlands, Dr. W. Junk Publishers, 137-144 p.

Jensen, D.L., Boddum, J.K., Tjell, J.C., and Christensen, T.H., 2002, The solubility of rhodochrosite (MnCO3) and siderite (FeCO3) in anaerobic aquatic environments: Applied Geochemistry, v. 17, p. 503-511.

Jensen, H.S., Kristensen, P., Jeppesen, E., and Skytthe, A., 1992, Iron: phosphorus ratio in surface sediment as an indicator of phosphate release from aerobic sediments in shallow lakes: Hydrobiologia, v. 235, p. 731-743.

JMP, 1999, Joint Monitoring Programme.

Johnes, P.J., 2007, Uncertainties in annual riverine phosphorus load estimation: Impact of load estimation methodology, sampling frequency, baseflow index and catchment population density: Journal of Hydrology, v. 332, p. 241-258.

Jones, M.J., 1985, The weathered zone aquifers of the basement-complex areas of Africa: Quarterly Journal of Engineering Geology, v. 18, p. 35-46.

Jönsson, H., Stintzing, A.R., Vinnerås, B., and Salomon, E., 2004, Guidelines on the Use of Urine and Faeces in Crop Production, EcoSanRes Report 2004-2: Stockholm, Sweden, The Stockholm Environment Institute.

Jordan, P., Arnscheidt, A., McGrogan, H., and McCormick, S., 2007, Characterising phosphorus transfers in rural catchments using a continuous bank-side analyser: Hydrology and Earth System Sciences, v. 11, p. 372-381.

Jordan, P., Arnscheidt, J., McGrogan, H., and McCormick, S., 2005, High-resolution phosphorus transfers at the catchment scale: the hidden importance of non-storm transfers: Hydrology and Earth System Sciences, v. 9, p. 685-691.

Kaggwa, R., 2009, National Water and Sewerage Coorporation (NWSC) Performance Improvement program, Uganda: Summary of achievements. Website: http://nwsc.co.ug/publications01.php (accessed 20th March 2009).

Kaiser, H.F., 1960, The application of electronic computers to factor analysis: Educational and psychological measurement, v. 20, p. 141-151.

Kansiime, F., Kateyo, E., Oryem-Origa, H., and Mucunguzi, P., 2007, Nutrient status and retention in pristine and disturbed wetlands in Uganda: Management implications. : Wetlands Ecology and Management, v. 15, p. 453-467.

Kansiime, F., and Nalubega, M., 1999a, Natural treatment by Uganda's Nakivubo swamp.: Water Quality International (MAR./APR.), p. 29-31.

Kansiime, F., and Nalubega, M., 1999b, Wastewater treatment by a natural wetland: the Nakivubo swamp, Uganda: processes and implications, AA Balkema Rotterdam, The Netherlands, 300 p.

Kansiime, F., Oryem-Origa, H., and Rukwago, S., 2005, Comparative assessment of the value of papyrus and cocoyams for the restoration of the Nakivubo wetland in Kampala, Uganda: Physics and Chemistry of the Earth, v. 30, p. 698-705.

Katukiza, A.Y., Ronteltap, M., Niwagaba, C.B., Foppen, J.W.A., Kansiime, F., and Lens, P.N.L., 2012, Sustainable sanitation technology options for urban slums: Biotechnology Advances, v. 30, p. 964-978.

Katukiza, A.Y., Ronteltap, M., Niwagaba, C.B., Kansiime, F., and Lens, P.N.L., 2014, A two-step crushed lava rock filter unit for grey water treatment at household level in an urban slum: Journal of Environmental Management, v. 133, p. 258-267.

Katukiza, A.Y., Ronteltap, M., Oleja, A., Niwagaba, C.B., Kansiime, F., and Lens, P.N.L., 2010a, Selection of sustainable sanitation technologies for urban slums - A case of Bwaise III in Kampala, Uganda: Science of the Total Environment, v. 409, p. 52-62.

Katukiza, A.Y., Ronteltap, M., Oleja, A., Niwagaba, C.B., Kansiime, F., and Lens, P.N.L., 2010b, Selection of sustainable sanitation technologies for urban slums — A case of Bwaise III in Kampala, Uganda: Science of the Total Environment, v. 409, p. 52-62.

KDMP, 2002, KDMP - Kampala Drainage Master Plan. Nakivubo Channel Rehabilitation Project, Ministry of Local Government, Kampala City Council, Uganda.

Kebede, S., Travi, Y., Alemayehu, T., and Ayenew, T., 2005, Groundwater recharge, circulation and geochemical evolution in the source region of the Blue Nile River, Ethiopia: Applied Geochemistry, v. 20, p. 1658-1676.

Kelbe, B.E., Bodenstein, B., and Mulder, G.J., 1991, Investigation of the hydrological response to informal settlements on small catchments in South Africa: IAHS Publication (International Association of Hydrological Sciences, p. 219-227.

Kelderman, P., Kansiime, F., Tola, M.A., and Van Dam, A., 2007, The role of sediments for phosphorus retention in the Kirinya wetland (Uganda). Wetlands Ecology and Management, v. 15, p. 481-488.

Kelderman, P., Koech, D.K., Gumbo, B., and O'Keeffe, J., 2009, Phosphorus budget in the low-income, peri-urban area of Kibera in Nairobi (Kenya): Water Sci Technol, v. 60, p. 2669-76.

Kemka, N., Njiné, T., Togouet, S.H.Z., Menbohan, S.F., Nola, M., Monkiedje, A., Niyitegeka, D., and Compère, P., 2006, Eutrophication of lakes in urbanized areas:

The case of Yaounde Municipal Lake in Cameroon, Central Africa. Lakes and Reservoirs: Research and Management, v. 11, p. 47-55.

Kemka, N., Togouet, S.H.Z., Kinfack, R.P.D., Nola, M., Menbohan, S.F., and Njine, T., 2009, Dynamic of phytoplankton size-class and photosynthetic activity in a tropical hypereutrophic lake: the Yaounde municipal lake (Cameroon): Hydrobiologia, v. 625, p. 91-103.

Kerven, G.L., Menzies, N.W., and Geyer, M.D., 2000, Soil carbon determination by high temperature combustion: A comparison with dichromate oxidation procedures and the influence of charcoal and carbonate carbon on the measured value: Communications in soil science and plant analysis, v. 31, p. 1935-1939.

Key, R., 1992, An Introduction to the Crystalline Basement of Africa, *in* Wright, E., and Burgess, W., eds., Hydrogeology of Crystalline Basement Aquifers in Africa, Volume Special Publication No. 66: London, Geological Society, p. 29-57

Kimani-Murage, E.W., and Ngindu, A.M., 2007, Quality of water the slum dwellers use: the case of a Kenyan slum: J Urban Health, v. 84, p. 829-38.

Kling, H.J., Mugidde, R., and Hecky, R.E., 2001, Recent changes in the phytoplankton community of Lake Victoria in response to eutrophication., *in* Munawar, M., and Becky, R., eds., The great lakes of the world (GLOW): food-web, health and integrity, Volume Ecovision World Monograph Series: Leiden, The Netherlands, Backhuys Publishers, p. 47-65.

Konert, M., and Vandenberghe, J.E.F., 1997, Comparison of laser grain size analysis with pipette and sieve analysis: a solution for the underestimation of the clay fraction: Sedimentology, v. 44, p. 523-535.

Kortatsi, B.K., 2006, Hydrochemical characterization of groundwater in the Accra plains of Ghana: Environmental Geology, v. 50 p. 299-311.

Kortatsi, B.K., 2007, Hydrochemical framework of groundwater in the Ankobra Basin, Ghana: Aquatic Geochemistry, v. 13, p. 41-74.

Kortatsi, B.K., Tay, C.K., Anornu, G., Hayford, E., and Dartey, G.A., 2008, Hydrogeochemical evaluation of groundwater in the lower Offin basin, Ghana: Environmental Geology, v. 53, p. 1651-1662.

Kulabako, N.R., Nalubega, M., and Thunvik, R., 2004, Characterisation of Peri-urban Anthropogenic Pollution in Kampala, Uganda, Proceedings of the 30th WEDC International conference on people centred approaches to water and environmental sanitation: 25-29th October. Vientiane: Lao PDR; 2004. p. 474–82.

Kulabako, N.R., Nalubega, M., and Thunvik, R., 2007, Study of the impact of land use and hydrogeological settings on the shallow groundwater quality in a peri-urban area of Kampala, Uganda: Science of the Total Environment, v. 381, p. 180-99.

Kulabako, N.R., Nalubega, M., and Thunvik, R., 2008, Phosphorus transport in shallow groundwater in peri-urban Kampala, Uganda: results from field and laboratory measurements: Environmental Geology, v. 53, p. 1535-1551.

Kulabako, N.R., Nalubega, M., Wozei, E., and Thunvik, R., 2010, Environmental health practices, constraints and possible interventions in peri-urban settlements in developing countries – a review of Kampala, Uganda: International Journal of Environmental Health Research, v. 20, p. 231-257.

Kulabako, N.R., Ssonko, N.K.M., and Kinobe, J., 2011, Greywater Characteristics and Reuse in Tower Gardens in Peri-Urban Areas-Experiences of Kawaala, Kampala, Uganda: Open Environmental Engineering Journal, v. 4, p. 147-154.

Kyambadde, J., Kansiime, F., and Dalhammar, G., 2005, Nitrogen and phosphorus removal in substrate-free pilot constructed wetlands with horizontal surface flow in Uganda: Water, Air, & Soil Pollution, v. 165, p. 37-59.

Lambrakis, N., Antonakos, A., and Panagopoulos, G., 2004, The use of multicomponent statistical analysis in hydrogeological environmental research: Water Research, v. 38, p. 1862-1872.

Lawrence, A.R., Gooddy, D.C., Kanatharana, P., Meesilp, W., and Ramnarong, V., 2000, Groundwater evolution beneath Hat Yai, a rapidly developing city in Thailand: Hydrogeology Journal, v. 8, p. 564-575.

Lawrence, A.R., Morris, B.L., Gooddy, D.C., Calow, R., and Bird, M.J., 1997, The study of the pollution risk to deep groundwater from urban waste waters: project summary report - Technical Report, WC/97/15: Keyworth, Nottingham, UK, British Geological Survey (BGS).

Love, D., Zingoni, E., Ravengai, S., Owen, R., Moyce, W., Mangeya, P., Meck, M., Musiwa, K., Amos, A., and Hoko, Z., 2006, Characterization of diffuse pollution of shallow groundwater in the Harare urban area, Zimbabwe, *in* Xu, Y., and Usher, B., eds., Groundwater pollution in Africa: Leiden, The Netherlands, Taylor & Francis/Balkema, p. 65-75.

Lutterodt, G., Foppen, J.W.A., and Uhlenbrook, S., 2014, Escherichia coli strains harvested from springs in Kampala, Uganda: cell characterization and transport in saturated porous media: Hydrological Processes, v. 28, p. 1973-1988.

Lyngkilde, J., and Christensen, T.H., 1992, Fate of organic contaminants in the redox zones of a landfill leachate pollution plume (Vejen, Denmark): Journal of Contaminant Hydrology, v. 10, p. 291-307.

Magadza, C.H.D., 2003, Lake Chivero: A Management Case Study. Lakes and Reservoirs: Research and Management, v. 8(2), p. 69-81.

Marsalek, J., Jimenez-Cisneros, B., Malmqvist, P., Goldenfum, J., and Chocat, B., 2008, Urban water cycle processes and interactions, Taylor and Francis, Leiden and UNESCO Publishing, Paris.

Marshall, B.E., and Falconer, A.C., 1973, Eutrophication of a Tropical African Impoundment (Lake McIlwaine, Rhodesia): Hydrobiologia, v. 43, p. 109-123.

Martynova, M., 2011, Fe/P concentration ratio in Mozhaisk reservoir deposits as an indicator of phosphate sorption: Water Resources, v. 38, p. 211-219.

Massmann, G., Pekdeger, A., and Merz, C., 2004, Redox processes in the Oderbruch polder groundwater flow system in Germany: Applied Geochemistry, v. 19, p. 863-886.

Matagi, S.V., 2002, Some issues of environmental concern in Kampala, the capital city of Uganda: Environmental Monitoring and Assessment, v. 77, p. 121-138.

Matsunaga, T., Karametaxas, G., Von Gunten, H., and Lichtner, P., 1993, Redox chemistry of iron and manganese minerals in river-recharged aquifers: A model interpretation of a column experiment: Geochimica et Cosmochimica Acta, v. 57, p. 1691-1704.

Menció, A., and Mas-Pla, J., 2008, Assessment by multivariate analysis of groundwater–surface water interactions in urbanized Mediterranean streams: Journal of Hydrology, v. 352, p. 355-366.

Mikac, N., Cosovic, B., Ahel, M., Andreis, S., and Toncic, Z., 1998, Assessment of groundwater contamination in the vicinity of a municipal solid waste landfill (Zagreb, Croatia): Water Science and Technology, v. 37, p. 37-44.

Mileham, L., Taylor, R., Thompson, J., Todd, M., and Tindimugaya, C., 2008, Impact of rainfall distribution on the parameterisation of a soil-moisture balance model of groundwater recharge in equatorial Africa: Journal of Hydrology, v. 359, p. 46-58.

Mireri, C., Atekyereza, P., Kyessi, A., and Mushi, N., 2007, Environmental risks of urban agriculture in the Lake Victoria drainage basin: A case of Kisumu municipality, Kenya: Habitat International, v. 31, p. 375-386.

Miret-Gaspa, M., 2004, Use of hydrochemical, microbiological and physical monitoring to determine contamination mechanisms of spring water discharging from a deeply weathered regolith aquifer (Kampala, Uganda): MSc thesis, University of Neuchâtel.

Mkandawire, T., 2008, Quality of groundwater from shallow wells of selected villages in Blantyre District, Malawi: Physics and Chemistry of the Earth, v. 33, p. 807-811.

Mladenov, N., Strzepek, K., and Serumola, O.M., 2005, Water quality assessment and modeling of an effluent-dominated stream, the Notwane River, Botswana: Environmental Monitoring and Assessment, v. 109, p. 97-121.

Monney, I., Odai, S.N., Buamah, R., Awuah, E., and Nyenje, P.M., 2013, Environmental impacts of wastewater from urban slums: case study - Old Fadama, Accra: International Journal of Development and Sustainability, v. 2, p. In Press.

Montangero, A., and Belevi, H., 2007, Assessing nutrient flows in septic tanks by eliciting expert judgement: a promising method in the context of developing countries: Water Res, v. 41, p. 1052-64.

Moore, R.D., 2004, Introduction to salt dilution gauging for streamflow measurement: Part 1: Streamline Watershed Management Bulletin, v. 7, p. 20-23.

Morris, B.L., Darling, W.G., Cronin, A.A., Rueedi, J., Whitehead, E.J., and Gooddy, D.C., 2006, Assessing the impact of modern recharge on a sandstone aquifer beneath a suburb of Doncaster, UK: Hydrogeology Journal, v. 14, p. 979-997.

Morris, B.L., Lawrence, A.R.L., Chilton, P.J.C., Adams, B., Calow, R.C., and Klinck, B.A., 2003, Groundwater and its susceptibility to degradation: a global assessment of the problem and options for management, Early warning and assessment report series, Volume RS. 03-3: Nairobi, Kenya, United Nations Environment Programme.

Moyo, N.A.G., and Worster, K., 1997, The Effects of Organic Pollution on the Mukuvisi River, Harare, Zimbabwe., *in* Moyo, N.A.G., ed., Lake Chivero: A Polluted Lake: Harare, Zimbabwe, University of Zimbabwe Publications, p. 53-63.

Mphepya, J., Galy-Lacaux, C., Lacaux, J., Held, G., and Pienaar, J., 2006, Precipitation Chemistry and Wet Deposition in Kruger National Park, South Africa: Journal of Atmospheric Chemistry, v. 53, p. 169-183.

Muggide, R., 1993, The increase in phytoplankton primary productivity and biomass in Lake Victoria (Uganda). : Verh. Int. Ver. Limnol., v. 25, p. 846-849.

Mugisha, P., Kansiime, F., Mucunguzi, P., and Kateyo, E., 2007, Wetland vegetation and nutrient retention in Nakivubo and Kirinya wetlands in the Lake Victoria basin of Uganda: Physics and Chemistry of the Earth, v. 32, p. 1359-1365.

Mukwaya, C., 2001, Assessment of groundwater dynamics and its vulnerability to pollution around Wobulenzi Town: An isotopic and hydrochemical study: Kampala Uganda, Unpublished MSc Thesis, Makerere University.

Munro, J.L., 1966, A Limnological Survey of Lake McLlwaine, Rhodesia.: Hydrobiologia, v. 28.

Murphy, J., and Riley, J.P., 1962, A modified single solution method for the determination of phosphate in natural waters: Analytica Chimica Acta, v. 27, p. 31-36.

Mwanuzi, F., Aalderink, H., and Mdamo, L., 2003, Simulation of pollution buffering capacity of wetlands fringing the Lake Victoria: Environment international, v. 29, p. 95-103.

MWE, 2007, Uganda water and sanitation sector performance report, Ministry of Water and Environment (MWE) - The Government of Uganda.

Natumanya, E., Kansiime, F., and Mwanuzi, F.L., 2010, Assessment of Nutrient Loading and Retention along Nsooba stream and Lubigi Wetland, Kampala, Uganda, Proceedings of 11th International WATERNET/WARFSA/GWP-SA Symposium: Victoria falls, Zimbabwe, 27-29 October 2010, p. 18.

Navarro, A., and Carbonell, M., 2007, Evaluation of groundwater contamination beneath an urban environment: The Besos river basin (Barcelona, Spain): Journal of Environmental Management, v. 85, p. 259-269.

Neal, C., Jarvie, H.P., Neal, M., Love, A.J., Hill, L., and Wickham, H., 2005, Water quality of treated sewage effluent in a rural area of the upper Thames Basin, southern England, and the impacts of such effluents on riverine phosphorus concentrations: Journal of Hydrology, v. 304, p. 103-117.

Nevondo, T.S., and Cloete, T.E., 1999, Bacterial and chemical quality of water supply in the Dertig village settlement: Water SA, v. 25, p. 215-220.

Nhapi, I., 2008, Inventory of water management practices in Harare, Zimbabwe: Water and Environment Journal, v. 22, p. 54-63.

Nhapi, I., Hoko, Z., Siebel, M.A., and Gijzen, H.J., 2002, Assessment of the major water and nutrient flows in the Chivero catchment area, Zimbabwe: Physics and Chemistry of the Earth, Parts A/B/C, v. 27, p. 783-792.

Nhapi, I., Siebel, M.A., and Gijzen, H.J., 2004, The impact of urbanisation on the water quality of Lake Chivero, Zimbabwe: Water and Environment Journal, v. 18, p. 44-49.

Nhapi, I., Siebel, M.A., and Gijzen, H.J., 2006, A proposal for managing wastewater in Harare, Zimbabwe: Water and Environment Journal, v. 20, p. 101-108.

Nhapi, I., and Tirivarombo, S., 2004, Sewage discharges and nutrient levels in Marimba River, Zimbabwe: Water SA, v. 30, p. 107-113.

Novozamsky, I., Lexmond, T.M., and Houba, V.J.G., 1993, A Single Extraction Procedure of Soil for Evaluation of Uptake of Some Heavy Metals by Plants: International Journal of Environmental Analytical Chemistry, v. 51, p. 47-58.

Nsubuga, F.B., Kansiime, F., and Okot-Okumu, J., 2004, Pollution of protected springs in relation to high and low density settlements in Kampala - Uganda: Physics and Chemistry of the Earth, v. 29, p. 1153-1159.

NWSC, 2008, Kampala Sanitation Program (KSP) - Feasibility study report: Kampala, Uganda, National Water and Sewerage cooperation (NWSC).

Nyarko, B.K., Kofi Essumang, D., Eghan, M.J., Reichert, B., van de Giesen, N., and Vlek, P., 2010, Use of isotopes to study floodplain wetland and river flow interaction in the White Volta River basin, Ghana: Isotopes in Environmental and Health Studies, v. 46, p. 91-106.

Nyenje, P.M., and Batelaan, O., 2009, Estimating effects of climate change on groundwater recharge and base flow in the upper Ssezibwa catchment, Uganda: Hydrol. Sci. J., v. 54, p. 713-726.

Nyenje, P.M., Foppen, J.W.A., Kulabako, R., Muwanga, A., and Uhlenbrook, S., 2013a, Nutrient pollution in shallow aquifers underlying pit latrines and domestic solid waste dumps in urban slums: Journal of Environmental Management, v. 122, p. 15-24.

Nyenje, P.M., Foppen, J.W.A., Uhlenbrook, S., Kulabako, R., and Muwanga, A., 2010, Eutrophication and nutrient release in urban areas of sub-Saharan Africa — A review: Science of the Total Environment, v. 408, p. 447-455.

Nyenje, P.M., Foppen, J.W.A., Uhlenbrook, S., and Lutterodt, G., 2013b, Using hydrochemical tracers to assess impacts of unsewered urban catchments on hydrochemistry and nutrients in groundwater: Hydrological Processes, p. In Press.

Nyenje, P.M., Meijer, L.M.G., Foppen, J.W.A., Kulabako, R., and Uhlenbrook, S., 2014, Phosphorus transport and retention in a channel draining an urban, tropical catchment with informal settlements: Hydrology and Earth System Sciences, v. 18, p. 1009-1025.

O'Shea, B., and Jankowski, J., 2006, Detecting subtle hydrochemical anomalies with multivariate statistics: an example from 'homogeneous' groundwaters in the Great Artesian Basin, Australia: Hydrological Processes, v. 20, p. 4317-4333.

Oberholster, P.J., Botha, A.-M., and Cloete, T.E., 2005, An overview of toxic freshwater cyanobacteria in South Africa with special reference to risk, impact and detection by molecular marker tools.: Biokem, v. 17, p. 57-71.

Oberholster, P.J., Botha, A.M., and Ashton, P.J., 2009, The influence of a toxic cyanobacterial bloom and water hydrology on algal populations and macroinvertebrate abundance in the upper littoral zone of Lake Krugersdrift, South Africa: Ecotoxicology, v. 18, p. 34-46.

Oberholster, P.J., Botha, A.M., and Cloete, T.E., 2008, Biological and chemical evaluation of sewage water pollution in the Rietvlei nature reserve wetland area, South Africa: Environmental Pollution, v. 156, p. 184-192.

Oguttu, H.W., Bugenyi, F.W.B., Leuenberger, H., Wolf, M., and Bachofen, R., 2008, Pollution menacing lake victoria: Quantification of point sources around Jinja Town, Uganda: Water Sa, v. 34, p. 89-98.

Okiror, G., Kansiime, F., Byamukama, D., and Banadda, E.N., 2009, The Variation Of Water Quality Along Lubigi Wetland, Proceedings 10th International WATERNET/WARFSA/GWPSA Symposium: 28 - 30 October 2009, Entebbe, Uganda.

Olli, G., Darracq, A., and Destouni, G., 2009, Field study of phosphorous transport and retention in drainage reaches: Journal of Hydrology, v. 365, p. 46-55.

Parkhurst, D.L., and Appelo, C.A.J., 1999, User's guide to PHREEQC (version 2) - A computer program for speciation, batch-reaction, one-dimensional transport, and inverse geochemical calculations, U.S. Geological Survey Water-Resources Investigations Report 99-4259, 312 p.

Parkhurst, D.L., Stollenwerk, K.G., and Colman, J.A., 2003, Reactive-Transport Simulation of Phosphorus in the Sewage Plume at the Massachusetts Military Reservation, Cape Cod, Massachusetts: U.S. GEOLOGICAL SURVEY v. Water-Resources Investigations Report 03-4017, Northborough, Massachusetts.

Peters, N.E., and Donohue, R., 2001, Nutrient transport to the Swan–Canning Estuary, Western Australia: Hydrological Processes, v. 15, p. 2555-2577.

Pettersen, J., and Hertwich, E.G., 2008, Critical review: life-cycle inventory procedures for long-term release of metals: Environmental Science & Technology, v. 42, p. 4639-4647.

Pieterse, N., Venterink, H.O., Schot, P., and Verkroost, A., 2005, Is nutrient contamination of groundwater causing eutrophication of groundwater-fed meadows?: Landscape ecology, v. 20, p. 743-753.

Pritchard, M., Mkandawire, T., and O'Neill, J.G., 2007, Biological, chemical and physical drinking water quality from shallow wells in Malawi: Case study of Blantyre, Chiradzulu and Mulanje: Physics and Chemistry of the Earth, v. 32, p. 1167-1177.

Pritchard, M., Mkandawire, T., and O'Neill, J.G., 2008, Assessment of groundwater quality in shallow wells within the southern districts of Malawi: Physics and Chemistry of the Earth, v. 33, p. 812-823.

Ptacek, C.J., 1998, Geochemistry of a septic-system plume in a coastal barrier bar, Point Pelee, Ontario, Canada: Journal of Contaminant Hydrology, v. 33, p. 293-312.

Raadgever, G.T., Mostert, E., and Giesen, N.C., 2012, Learning from Collaborative Research in Water Management Practice: Water resources management, v. 26, p. 3251-3266.

Rauret, G., 1998, Extraction procedures for the determination of heavy metals in contaminated soil and sediment: Talanta, v. 46, p. 449-455.

Reddy, K.R., Kadlec, R.H., Flaig, E., and Gale, P.M., 1999, Phosphorus Retention in Streams and Wetlands: A Review: Critical Reviews in Environmental Science and Technology, v. 29, p. 83-146.

Richardson, C.J., and Vaithiyanathan, P., 1995, Phosphorus sorption characteristics of Everglades soils along a eutrophication gradient: Soil Science Society of America Journal, v. 59, p. 1782-1788.

Rivett, M.O., Buss, S.R., Morgan, P., Smith, J.W., and Bemment, C.D., 2008, Nitrate attenuation in groundwater: a review of biogeochemical controlling processes: Water Research, v. 42, p. 4215-4232.

Robarts, R.D., and Southall, G.C., 1977, Nutrient Limitation of Phytoplankton Growth in Seven Tropical Man-Made Lakes, with Specific Reference to Lake McIlwaine, Rhodesia.: Hydrobiologia, v. 79, p. 1-35.

Robertson, W.D., 2008, Irreversible phosphorus sorption in septic system plumes?: Ground Water, v. 46, p. 51-60.

Robertson, W.D., Moore, T.A., Spoelstra, J., Li, L., Elgood, R.J., Clark, I.D., Schiff, S.L., Aravena, R., and Neufeld, J.D., 2012, Natural attenuation of septic system nitrogen by anammox: Ground Water, v. 50, p. 541-53.

Robertson, W.D., Schiff, S.L., and Ptacek, C.J., 1998, Review of Phosphate Mobility and Persistence in 10 Septic System Plumes: Ground Water, v. 36, p. 1000-1010.

Rodríguez-Blanco, M.L., Taboada-Castro, M.M., and Taboada-Castro, M.T., 2013, Phosphorus transport into a stream draining from a mixed land use catchment in Galicia (NW Spain): Significance of runoff events: Journal of Hydrology, v. 481, p. 12-21.

Ruban, V., López-Sánchez, J.F., Pardo, P., Rauret, G., Muntau, H., and Quevauviller, P., 2001, Development of a harmonised phosphorus extraction procedure and certification of a sediment reference material: Journal of environmental monitoring, v. 3, p. 121-125.

Runkel, R.L., and Bencala, K.E., 1995, Transport of reacting solutes in rivers and streams, *in* Runkel, R.L., ed., Environmental Hydrology: Dordrecht, The Netherlands, Kluwer Academic, p. 137–164.

Salifu, A., Petrusevski, B., Ghebremichael, K., Buamah, R., and Amy, G., 2012, Multivariate statistical analysis for fluoride occurrence in groundwater in the Northern region of Ghana: Journal of Contaminant Hydrology, v. 140–141, p. 34-44.

Scheren, P.A.G.M., Zanting, H.A., and Lemmens, A.M.C., 2000, Estimation of water pollution sources in Lake Victoria, East Africa: Application and elaboration of the rapid assessment methodology: Journal of Environmental Management, v. 58, p. 235-248.

Schilling, K.E., 2002, Occurrence and Distribution of Ammonium in Iowa Groundwater: Water Environment Research, v. 74, p. 177-186.

Schollenberger, C.J., and Simon, R.H., 1945, Determination of Exchange Capacity and Exchangeable Bases in Soil-Ammonium Acetate Method: Soil science, v. 59, p. 13-24.

Séguis, L., Kamagaté, B., Favreau, G., Descloitres, M., Seidel, J.-L., Galle, S., Peugeot, C., Gosset, M., Le Barbé, L., and Malinur, F., 2011, Origins of streamflow in a crystalline basement catchment in a sub-humid Sudanian zone: The Donga basin (Benin, West Africa): Inter-annual variability of water budget: Journal of Hydrology, v. 402, p. 1-13.

Shenker, M., Seitelbach, S., Brand, S., Haim, A., and Litaor, M.I., 2005, Redox reactions and phosphorus release in re-flooded soils of an altered wetland: European Journal of Soil Science, v. 56, p. 515-525.

Shewa, W., and Geleta, B., 2010, Greywater tower Arba Minch, Ethiopia - Case study of sustainable sanitation projects. Sustainable Sanitation Alliance (SuSanA). Web: http://www.susana.org/lang-en/case-studies?view=ccbktypeitem&type=2&id=90 (Accessed March 15, 2014).

Smith, V.H., Tilman, G.D., and Nekola, J.C., 1999, Eutrophication: impacts of excess nutrient inputs on freshwater, marine, and terrestrial ecosystems: Environmental Pollution, v. 100, p. 179-196.

Søndergaard, M., Jensen, J., and Jeppesen, E., 1999, Internal phosphorus loading in shallow Danish lakes: Hydrobiologia, v. 408-409, p. 145-152.

Spiteri, C., Slomp, C.P., Regnier, P., Meile, C., and Van Cappellen, P., 2007, Modelling the geochemical fate and transport of wastewater-derived phosphorus in contrasting groundwater systems: Journal of Contaminant Hydrology, v. 92, p. 87-108.

Stephenson, D., 1991, Effects of urbanisation on catchment water balance, Report 183/12/93: Johannesburg, South Africa, Report 183/12/93, Water Systems Research Group, University of Witwatersrand.

Stevenson, F.J., 1986, Cycles of soil : carbon, nitrogen, phosphorus, sulfur, micronutrients: New York, John Wiley and Sons, 380 p.

Stigter, T., Carvalho Dill, A., and Ribeiro, L., 2011, Major issues regarding the efficiency of monitoring programs for nitrate contaminated groundwater: Environmental Science & Technology, v. 45, p. 8674-8682.

Stumm, W., and Morgan, J., 1981, Aquatic Chemistry: Wiley-Interscience, New York, p. 780.

Stutter, M.I., Langan, S.J., and Cooper, R.J., 2008, Spatial contributions of diffuse inputs and within-channel processes to the form of stream water phosphorus over storm events: Journal of Hydrology, v. 350, p. 203-214.

Stuyfzand, P.J., 1989, A new hydrochemical classification of watertypes: IAHS Publ, v. 182, p. 89-98.

Stuyfzand, P.J., 1993, Hydrochemistry and hydrology of the coastal dune area of the Western Netherlands. Ph.D Thesis Vrije Univ. Amsterdam: KIWA, p. 366.

Talling, J.F., 1963, Origin of stratification in an African rift lake.: Limnol. Oceanogr., v. 8, p. 68-78.

Talling, J.F., 1992, Environmental regulation in African shallow lakes and wetlands.: Rev. hydrobiol. trop., v. 25, p. 87-144.

Talling, J.F., and Talling, I.B., 1965, The photosynthetic activity of phytoplankton in East African lakes.: Int. Rev. Gesamten Hydrobiol., v. 50, p. 1-32.

Taylor, R.G., Cronin, A.A., Lerner, D.N., Tellam, J.H., Bottrell, S.H., Rueedi, J., and Barrett, M.H., 2006, Hydrochemical evidence of the depth of penetration of anthropogenic recharge in sandstone aquifers underlying two mature cities in the UK: Applied Geochemistry, v. 21, p. 1570-1592.

Taylor, R.G., and Howard, K.W.F., 1996, Groundwater recharge in the Victoria Nile basin of east Africa: Support for the soil moisture balance approach using stable isotope tracers and flow modelling: Journal of Hydrology, v. 180, p. 31-53.

Taylor, R.G., and Howard, K.W.F., 1998, The dynamics of groundwater flow in the regolith of Uganda, *in* Dillon, P., and Simmers, I., eds., Shallow Groundwater Systems: Rotterdam, Balkema, p. 97 - 113.

Taylor, R.G., and Howard, K.W.F., 1999a, The influence of tectonic setting on the hydrological characteristics of deeply weathered terrains: evidence from Uganda: Journal of Hydrology, v. 218, p. 44-71.

Taylor, R.G., and Howard, K.W.F., 1999b, Lithological evidence for the evolution of weathered mantles in Uganda by tectonically controlled cycles of deep weathering and stripping: Catena, v. 35, p. 65-94.

Taylor, R.G., Miret-Gaspa, M., Tumwine, J., Mileham, L., Flynn, R., Howard, G., and Kulabako, R., 2009, Increased risk of diarrhoeal diseases from climate change: evidence from communities supplied by groundwater in Uganda, *in* Taylor, R., Tindimugaya, C., M., O., and Shamsudduha, M., eds., Groundwater and Climate in Africa, Proceedings of the Kampala Conference, International Association of Hydrological Sciences (IAHS) Publication 334, p. 15-19.

Thornton, J.A., 1989, Aspects of the phosphorus cycle in Hartbeespoort Dam (South Africa): Hydrobiologia, v. 183, p. 87-95.

Thornton, J.A., and Ashton, P.J., 1989, Aspects of the phosphorus cycle in Hartbeespoort Dam (South Africa) Phosphorus loading and seasonal distribution of phosphorus in the reservoir: Hydrobiologia, v. 183, p. 73-85.

Thornton, J.A., and Nduku, W.K., 1982, Water chemistry and nutrient budgets, Lake Mcilwaine, Springer, p. 43-59.

Thornton, J.A., Rast, W., Holland, M.M., Jolankai, G., and Ryding, S.-O., 1999, Assessment and control of nonpoint source pollution of aquatic ecosystems: a practical approach: New York, USA, UNESCO, Parthenon Publishing Group, xi, 466 p.

Tournoud, M.G., Perrin, J.L., Gimbert, F., and Picot, B., 2005, Spatial evolution of nitrogen and phosphorus loads along a small Mediterranean river: implication of bed sediments: Hydrological Processes, v. 19, p. 3581-3592.

Tredoux, G., and Talma, A.S., 2006, Nitrate pollution of groundwater in southern Africa, *in* Xu, Y., and Usher, B., eds., Groundwater pollution in Africa: Leiden, The Netherlands, Taylor & Francis/Balkema, p. 15-36.

Trudell, M.R., Gillham, R.W., and Cherry, J.A., 1986, An in-situ study of the occurrence and rate of denitrification in a shallow unconfined sand aquifer: Journal of Hydrology, v. 83, p. 251-268.

Uma, K.O., 1993, Nitrates in Shallow (Regolith) Aquifers around Sokoto Town, Nigeria: Environmental Geology, v. 21, p. 70-76.

UN-Habitat, 2003, The challenge of slums: global report on human settlements 2003: London, Earthscan Publication Ltd.

UN-Habitat, 2008, The State of African Cities - A framework for addressing urban challenges in Africa: Nairobi, Kenya, United Nations Human Settlements Programme (UN-HABITAT), p. 207.

van Breukelen, B.M., and Griffioen, J., 2004, Biogeochemical processes at the fringe of a landfill leachate pollution plume: potential for dissolved organic carbon, Fe(II), Mn(II), NH4, and CH4 oxidation: Journal of Contaminant Hydrology, v. 73, p. 181-205.

van Dam, A.A., Dardona, A., Kelderman, P., and Kansiime, F., 2007, A simulation model for nitrogen retention in a papyrus wetland near Lake Victoria, Uganda (East Africa): Wetlands Ecology and Management, v. 15, p. 469-480.

Verhoeven, J.T., Koerselman, W., and Meuleman, A.F., 1996, Nitrogen- or phosphorus-limited growth in herbaceous, wet vegetation: relations with atmospheric inputs and management regimes: Trends Ecol Evol, v. 11, p. 494-7.

Verschuren, D., Johnson, T.C., Kling, H.J., Edgington, D.N., Leavitt, P.R., Brown, E.T., Talbot, M.R., and Hecky, R.E., 2002, History and timing of human impact on Lake Victoria, East Africa: Proceedings of the Royal Society B-Biological Sciences, v. 269, p. 289-294.

Viner, A.B., Breen, C., Golterman, H.L., and Thornton, J.A., 1981, Nutrient budgets, *in* Symoens, J.J., Burgis, M., and Gaudet, J.J., eds., The ecology and utilization of African inland waters, p. 137–48.

von Sperling, M., and de Lemos Chernicharo, C.A., 2005, Biological Wastewater Treatment in Warm Climate Regions: London, UK, IWA Publishing.

Vos, A.T., and Roos, J.C., 2005, Causes and consequences of algal blooms in Loch Logan, an urban impoundment: Water Sa, v. 31, p. 385-392.

Wakida, F.T., and Lerner, D.N., 2005, Non-agricultural sources of groundwater nitrate: a review and case study: Water Research, v. 39, p. 3-16.

Walkley, A., and Black, I.A., 1934, An examination of the Degtjareff method for determining soil organic matter, and a proposed modification of the chromic acid titration method: Soil science, v. 37, p. 29-38.

Webb, B.W., Phillips, J.M., Walling, D.E., Littlewood, I.G., Watts, C.D., and Leeks, G.J.L., 1997, Load estimation methodologies for British rivers and their relevance to the LOIS RACS(R) programme: Science of the Total Environment, v. 194, p. 379-389.

Weiskel, P.K., and Howes, B.L., 1992, Differential transport of sewage-derived nitrogen and phosphorus through a coastal watershed: Environmental Science & Technology, v. 26, p. 352-360.

WHO, 2000, Water Supply and Sanitation Sector Report Year 2000 - Africa Regional Assessment., World Health Organization.

WHO, and UNICEF, 2012, Progress on Drinking Water and Sanitation: 2012 Update. WHO/UNICEF Joint Monitoring Programme for Water Supply and Sanitation (JMP), Joint Monitoring Programme for Water Supply and Sanitation (JMP).

WHO/UNICEF, 2013, Progress on sanitation and drinking-water - 2013 update, World Health Organization, Geneva.

Wiegand, C., and Pflugmacher, S., 2005, Ecotoxicological effects of selected cyanobacterial secondary metabolites a short review: Toxicology and Applied Pharmacology, v. 203, p. 201-218.

Withers, P.J.A., Jarvie, H.P., and Stoate, C., 2011, Quantifying the impact of septic tank systems on eutrophication risk in rural headwaters: Environment International, v. 37, p. 644-653.

Witte, F., Welten, M., Heemskerk, M., van der Stap, I., Ham, L., Rutjes, H., and Wanink, J., 2008, Major morphological changes in a Lake Victoria cichlid fish within two decades: Biological Journal of the Linnean Society, v. 94, p. 41-52.

Wolf, L., Morris, B., and Burn, S., 2006, AISUWRS: Urban Water Resources Toolbox-Integrating Groundwater into Urban Water Management: London, UK, IWA Publishing.

Wood, W.W., and Sanford, W.E., 1995, Chemical and isotopic methods for quantifying ground-water recharge in a regional, semiarid environment: Ground Water, v. 33, p. 458-468.

Wu, D., Ekama, G.A., Wang, H.-G., Wei, L., Lu, H., Chui, H.-K., Liu, W.-T., Brdjanovic, D., van Loosdrecht, M., and Chen, G.-H., 2014, Simultaneous nitrogen and phosphorus removal in the sulfur cycle-associated Enhanced Biological Phosphorus Removal (EBPR) process: Water Research, v. 49, p. 251-264.

WWAP, 2009, The United Nations World Water Development Report 3: Water in a Changing World: World Water Assessment Programme (WWAP), Paris: UNESCO, and London, Earthscan.

Xu, Y., and Usher, B., 2006, Issues of groundwater pollution in Africa, *in* Xu, Y., and Usher, B., eds., Groundwater pollution in Africa: Leiden, The Netherlands, p. 3-9.

Yang, S.Y., and Yeh, H.D., 2004, A simple approach using Bouwer and Rice's method for slug test data analysis: Ground Water, v. 42, p. 781-4.

Yidana, S.M., Banoeng-Yakubo, B., Akabzaa, T., and Asiedu, D., 2011, Characterization of the groundwater flow regime and hydrochemistry of groundwater from the Buem formation, Eastern Ghana: Hydrological Processes, v. 25, p. 2288-2301.

Yidana, S.M., Ophori, D., and Banoeng-Yakubo, B., 2008, Hydrogeological and hydrochemical characterization of the Voltaian Basin: The Afram Plains area, Ghana: Environmental Geology, v. 53, p. 1213-1223.

Zanini, L., Robertson, W.D., Ptacek, C.J., S.L., S., and T, M., 1998, Phosphorus characterization in sediments impacted by septic effluent at four sites in central Canada: Journal of Contaminant Hydrology, v. 33, p. 405–429.

Zhang, Z., Fukushima, T., Onda, Y., Gomi, T., Fukuyama, T., Sidle, R., Kosugi, K., and Matsushige, K., 2007, Nutrient runoff from forested watersheds in central Japan during typhoon storms: implications for understanding runoff mechanisms during storm events: Hydrological Processes, v. 21, p. 1167-1178.

Zilberg, B., 1966, Gastro-enteritis in Salisbury European children: a five-year study.: Central African Journal of Medicine, v. 12(9), p. 164-168.

Zinabu, G.M., Kebede-Westhead, E., and Desta, Z., 2002, Long-term changes in chemical features of waters of seven Ethiopian rift-valley lakes: Hydrobiologia, v. 477, p. 81-91.

Zinabu, G.M., and Taylor, W.D., 1989, Seasonal and spatial variation in abundance, biomass and activity of heterotrophic bacterioplankton in relation to some biotic and abiotic variables in an Ethiopian rift-valley lake (Awassa). Freshwat. Biol., v. 22, p. 355-368.

Zingoni, E., Love, D., Magadza, C., Moyce, W., and Musiwa, K., 2005, Effects of a semi-formal urban settlement on groundwater quality Epworth (Zimbabwe): Case study and groundwater quality zoning: Physics and Chemistry of the Earth, v. 30, p. 680-688.

Zunckel, M., Robertson, L., Tyson, P.D., and Rodhe, H., 2000, Modelled transport and deposition of sulphur over Southern Africa: Atmospheric Environment, v. 34, p. 2797-2808.

Zurawsky, M.A., Robertson, W.D., Ptacek, C.J., and Schiff, S.L., 2004, Geochemical stability of phosphorus solids below septic system infiltration beds: Journal of Contaminant Hydrology, v. 73, p. 129-143.

About the Author

Philip Nyenje was born on 28 August 1977 in Mengo - Kampala, Uganda. He joined Makerere University in September 1998 and graduated with a Bachelor of Science degree in Civil Engineering with a first class in January 2003. Thereafter, he worked as a project engineer with the Ministry of Works in a country-wide investment program to design and upgrade of district roads in Uganda. In 2004, he joined the Ministry of Water and Environment as a water engineer. Here, he was attached to a project aimed at improving safe water supply and sanitation in growing towns in south-western Uganda. It is  from here that Philip's interest in water research grew particularly, in the need to access safe and clean water resources as well as the protection of these resources from pollution and global changes. In September 2005, Philip was awarded a VLIR scholarship to pursue an MSc degree in Water Resources Engineering at the Katholieke Universiteit of Leuven and the Vrije Universiteit of Brussels in Belgium. He graduated in 2007 with greatest distinction and his thesis was entitled 'Estimating the effect of climate change on water resources in Uganda'. Thereafter, he worked as a research assistant to develop a new module for Water Resources Engineering program of the Katholieke University of Leuven. Philip then started his PhD research in January 2009 and he has been focusing on understanding the mechanisms governing the transport and fate of sanitation-related nutrients in groundwater and surface water in informal settlements (or slums) in mega-cities in sub-Saharan Africa. During his PhD research, Philip followed the Educational Programme of SENSE (Socio - Economic and Natural Sciences of the Environment) and obtained a certificate in September 2014. He has presented his work at several international conferences and has published several papers in highly-ranked international peer-reviewed journals.

List of publications related to this study:

Nyenje, P.M., Meijer, L.M.G., Foppen, J.W., Kulabako, R., and Uhlenbrook, S., 2014, Phosphorus transport and retention in a channel draining an urban, tropical catchment with informal settlements: Hydrology and Earth System Sciences, v. 18, p. 1009-1025.

Nyenje, P.M., Havik, J.C.N., Foppen, J.W., Muwanga, A. and Kulabako, R. 2013, Understanding the fate of sanitation-related nutrients in a shallow sandy aquifer below an urban slum area: Journal of Contaminant Hydrology, http://dx.doi.org/10.1016/j.jconhyd.2014.06.011."In Press"

Nyenje, P. M., Meijer, L. M. G., Foppen, J. W., Kulabako, R., & Uhlenbrook, S., 2013c. Transport and retention of phosphorus in surface water in an urban slum area: Hydrology and Earth System Sciences Discussions, 10(8), 10277-10312.

Nyenje, P. M., Foppen, J. W, Uhlenbrook, S & Lutterodt, G., 2013b, Using hydro-chemical tracers to assess impacts of unsewered urban catchments on hydrochemistry and nutrients in groundwater: Hydrological processes. DOI: 10.1002/hyp.10070.

Nyenje, P. M., Foppen, J. W., Kulabako, R., Muwanga, A., & Uhlenbrook, S., 2013a, Nutrient pollution in shallow aquifers underlying pit latrines and domestic solid waste dumps in urban slums: Journal of Environmental Management, 122, 15-24.

Nyenje, P. M., Foppen, J. W., Uhlenbrook, S., Kulabako, R., & Muwanga, A., 2010, Eutrophication and nutrient release in urban areas of sub-Saharan Africa—a review. Science of the Total Environment, 408(3), 447-455.

Monney, I., Buamah, R., Odai, S. N., Awuah, E., & Nyenje, P. M., 2013, Evaluating Access to Potable Water and Basic Sanitation in Ghana's Largest Urban Slum Community: Old Fadama, Accra: Journal of Environment and Earth Science, 3(11), 72-79.

Netherlands Research School for the
Socio-Economic and Natural Sciences of the Environment

C E R T I F I C A T E

The Netherlands Research School for the
Socio-Economic and Natural Sciences of the Environment
(SENSE), declares that

Philip Mayanja Nyenje

born on 28 August 1977 in Kampala, Uganda

has successfully fulfilled all requirements of the
Educational Programme of SENSE.

Delft, 15 September 2014

the Chairman of the SENSE board   the SENSE Director of Education

Prof. dr. Rik Leemans   Dr. Ad van Dommelen

The SENSE Research School has been accredited by the Royal Netherlands Academy of Arts and Sciences (KNAW)

K O N I N K L I J K E   N E D E R L A N D S E
A K A D E M I E   V A N   W E T E N S C H A P P E N

The SENSE Research School declares that Mr. Philip Mayanja Nyenje has successfully fulfilled all requirements of the Educational PhD Programme of SENSE with a work load of 44.4 ECTS, including the following activities:

SENSE PhD Courses
- Environmental Research in Context
- Research Context Activity: Dissemination of PhD Research findings on Sanitation in Slums: organizing stakeholder meeting, publishing newspaper article, and producing two instructive movies
- SENSE Writing Week

Other PhD and Advanced MSc Courses
- Introduction to Tracer Hydrology
- Sanitation-related urban groundwater pollution
- Techniques for Writing and Presenting a Scientific Paper
- Project and Time Management (P&TM)

Management and Didactic Skills Training
- Supervision of MSc Theses:
    - 'The fate and transport of nutrients in shallow groundwater and soil of an urban slum area in the city of Kampala, Uganda'
    - 'Understanding the fate and transport of phosphorus in streams draining an unsewered slum in sub-Saharan Africa'

Oral Presentations
- *Hydro-chemical characterization of groundwater and surface water in urban slum dominated catchments in sub-Saharan Africa.* 11th WaterNet/WARFSA/GWP-SA Symposium, Victoria falls, Zimbabwe, 27-29 October 2010, Victoria falls, Zimbabwe
- *Hydrological aspects of improved sanitation in urban slums.* WaterNET Conference 2009, Entebbe Uganda, 28-30 October 2009, Entebbe, Uganda
- *Understanding the fate of nutrients in groundwater and surface water draining unsewered urban slums in sub-Saharan Africa.* Boussinesq Hydrology Lecture 2011, Royal Netherlands Academy of Arts and Sciences (KNAW), 20 October 2011, Amsterdam, Netherlands

SENSE Coordinator PhD Education

Dr. ing. Monique Gulickx